geodesic math

and how to use it

UNIVERSITY OF CALIFORNIA PRESS BERKELEY • LOS ANGELES • LONDON

geodesic math

and how to use it

by Hugh Kenner

University of California Press
Berkeley and Los Angeles, California

University of California Press, Ltd.
London, England

ISBN 0-520-02924-0
Library of Congress Catalog Card Number: 74-27292

Printed in the United States of America

CONTENTS

What This Book Is

Geodesics is a technique for making shell-like structures that hold
themselves up without supporting columns, by exploiting a three-
way grid of tensile forces. They are very strong and can be very
large: the geodesic bubble erected to house the United States
exhibits at Expo '67 in Montreal encloses some 6 million cubic
feet; it is approximately three-fourths of a sphere 250 feet in
diameter. They are also very light for what they do: the Montreal
bubble weighs about 600 tons, Plexiglas skin and all. It has been
calculated that a geodesic sphere approximately one-half mile in
diameter would float away like a soap bubble if the air inside it
were one degree warmer than the air outside. They have been
used as homes, as offices, as fair pavilions, as locomotive round-
houses, as gymnasiums, as auditoriums, as banks, as playground
structures for children to climb on, as housings for radar installa-
tions on the DEW line, and for observatories buried under snow at
the South Pole. Yet, considering their apparent potential, in the
quarter-century since Buckminster Fuller introduced them they
haven't been used very widely.

That is partly because they are mathematically derived structures, and the mathematics hasn't been easily available. Parts for
the self-supporting frame must be fabricated to close specifications. The fabrication, with today's technology, is no problem; the
problem is learning what the specifications should be. If we know
them, we can achieve extraordinary savings of material, weight,
and effort. If we don't, we have no resource save one of the
conventional methods of building, which by geodesic standards
means gross overbuilding.

My assumption is that if architects, designers, engineers knew
how to get past the first step, which is calculating the pertinent
details of a geodesic structure's geometry, they would explore
geodesic potentials more than they have. This book shows how to

do that — and without access to a large computer. Hobbyists and math buffs should find it interesting, too.

It would help, moreover, if geodesic design could be disentangled from its historical reliance on spheres. More than once a geodesic approach has been shunned because the designer, for one good reason or for several, didn't want his structure to resemble a slice from a sphere. But anyone who masters the design procedure in this book will find that, once he has mastered geodesic spheres, geodesic eggs, whether tall or flat, are only one step more complicated, and free-form contours are no more difficult than eggs. You can work from an equation if you have one; or you can even draw a contour freehand and calculate without much trouble an equivalent geodesic structure.

Finally, the great strength of geodesic structures needs demystifying. They are always much lighter and tougher than people expect, partly because we are used to thinking that walls bear weight the way posts do, and are unprepared for the resources of a hidden tensile system. In fact, Fuller's geodesic domes constitute a special case of a larger class of Fuller constructs called Tensegrities, and the way to an intuitive understanding of domes is to understand Tensegrity first. So the book begins by developing Tensegrity mathematics, something hitherto unexplored. (That may be one reason no useful structures exploiting pure Tensegrity — tension wholly separated from compression — have been built at all.)

As to what you can do with the calculations: you can immediately build fascinating models, out of sticks and string and soda straws and paper fasteners. If you want to build something useful to get inside of, you will be well advised to build a model first. After that you will encounter other problems for which I have no help to offer. I can show you how to calculate, to any degree of accuracy you need, the length of every member in the framing system, and every angle governing their junctions. You yourself will have to decide what to make the members from, how to join them, how to apply a skin made out of what, how to waterproof and fenestrate and ventilate the structure, what to do with the space inside. All this entails a whole new building technology which, despite some spectacular achievements, may be estimated to be still in the log-cabin stage. If inability to make the first calculations has helped ensure that people who might be solving such problems are employing their time otherwise, then this book may help unblock development.

There is no use anyone's pretending that there are no problems. There are problems in keeping an ordinary flat roof from leaking, problems generations of building experience have pretty well solved. When someone nails plywood triangles onto a wooden geodesic frame and discovers that cold weather pulls the joints apart, he has no more discredited geodesics than a leaky outhouse discredits the procedures that will ensure a snug bungalow. He has merely proved that one kind of geodesic technology won't work. (Fiberglassing the whole exterior of a wooden dome apparently will work, and may be the cheapest way to get a homogenous shell. Or there may be better ways.)

Strength calculations remain another problem area. Many geodesic structures — including Fuller's own first commercial building, the dome over the Ford rotunda in Dearborn, Michigan — have been overbuilt because no one really knew what the minimum framing system would be. That hidden tensile net defeats post-and-beam calculations. Shell analysis — which in effect postulates an infinity of components of zero dimension — is said to yield the most reliable results, but one would think something less distant from structural reality could be devised. Possibly this book's approach via Tensegrity may suggest to some imaginative engineer the way to go.

The Tensegrities themselves are tantalizing. Everyone who sees a model of one — resilient, nearly indestructible despite its local fragilities, recovering like a rubber ball from gross deformations — thinks at once that a building framed like that would be virtually invulnerable to earthquakes. So far as I know, no building has ever been framed like that: the skin, for one thing, would need to be as flexible as the frame. Fuller at one time intended that the Montreal bubble should be a giant Tensegrity, but time and budget inhibited the necessary research. If Tensegrity has a practical use, other than yielding models and pieces of sculpture and helping us understand geodesics, the first principles of that usefulness remain to be investigated. By making a rigorous design procedure available, this book may get some visionary genius started. I trust tests will be run before any contracts are let. I have absolutely no idea how any of these designs would behave on a large scale under load. I make no claims whatever in that domain and decline all responsibility for misfortune.

I am neither an engineer nor a mathematician. I've checked most things out with small models, but that has been the extent of my hardware implementation. I've thought it best to assume that the reader's mathematics is about where mine was when Bucky Fuller's proximity got me interested in this subject: algebra and high-school trig, inoperative from decades of disuse. I have also had the great advantage of *not* having access to a large digital computer. Computers in the past have helped obfuscate the subject by making unsuitable methods workable. It was in trying to avoid the sheer donkeywork a computer shoulders uncomplainingly that I stumbled on the applicability of the spherical coordinate system, which I came across in Professor George E. Owen's *Fundamentals of Scientific Mathematics* (Baltimore, Johns Hopkins Press: 1961).

To avoid mind-numbing weariness, you do need a good pocket calculator with trig and power functions. I used the Hewlett-Packard HP-35, which can handle any procedure in this book without one's having to write down intermediate results. For the more cumbersome equations I give HP-35 and HP-21 routines here. Users of other calculators will have to devise their own.

For repetitive work — and much work on large-scale geodesic structures is repetitive — the programmable HP-65 is of course still better. All you do is enter data; it repeats routines indefinitely without chance of error. I got one when I started compiling the tables. But the simpler machine works fine.

A word for mathematicians. Conceptually this book is very simple. Part 1 offers an essentially Newtonian analysis of Tensegrities, treating them as diagrams of their own system of equilibrated forces, and shows how they can be designed by using trig to predict their equilibrium states. There is also a brief account of an approach using calculus, which depends on the fact that the tension system seeks a minimum length. Part 2 simply applies in detail the elementary insight that spherical structures are best described by spherical coordinates. Books explain this coordinate system as a sort of conceptual metaphor but don't seem to explain how to use it, so I've had to devise most of the algorithms myself. They may not be the most elegant possible, but they all work.

At three points where my own resources failed me I received invaluable help from Professor Paul Kelly, University of California, Santa Barbara; Mr. Roland O. Davis of Goleta, California; and Professors George Owen and Rufus Isaacs at The Johns Hopkins University.

A letter from Professor H. S. M. Coxeter, explaining a tantalizing regularity I'd noticed, gave the key to a simple method of finding the coordinates themselves; Mrs. Cindy Engers first glimpsed the way of putting it into workable form.

Much of the book's terminology is that of Joe Clinton, whose *Advanced Structural Design Concepts for Future Space Missions* (NASA Contract NGR-14-008-002, 1970) remains the pioneer job of systematizing geodesic geometry. I am grateful to him for a copy. When the definitive treatment of geodesics is compiled, he will be at the center of the enterprise. Mine is meanwhile an interim report, to go on from. Peter Calthorpe, Lloyd Kahn, and Bob Easton were generous with help when I was getting started. All future workers will be indebted to Kahn's *Domebook 1* (Los Gatos, Ca., Pacific Domes: 1970) and *Domebook 2* (Bolinas, Ca., Pacific Domes: 1971) for first making information accessible. And historians will some day disentangle from the early history of the art the contributions of Donald Richter and Duncan Stuart, neither of whom I've met. Much of the time I've been aware of handling ideas — I don't know which ones — that they and others first formulated; just as much of my analysis of Tensegrity was incited by structures Tony Pugh first showed me, which in turn embodied concepts Kenneth Snelson — unknown to Tony — had once intuited. Everyone in this list is indebted — as am I — to Buckminster Fuller, instigator nonpareil.

Part One

tensegrity

1. Weight vs. Tension

Beams will support a roof, and an easy way to support the beams
is to put posts under them. One drawback is that the posts clutter
up your floor space. You can line posts up and enclose them in a
partition wall and pretend you wanted the wall there anyway,
whether you did or not. What you can't do, without recourse to
clever engineering, is free up floor space by moving the posts wide
apart. If you do that, the beam starts to sag in the middle. If it
sags too much, it breaks.

The limit in spacing posts is set by the material of the beam, as
the builders of Stonehenge apparently understood (Diagram 1.1).
If the crossbeam sags, its upper edge will be compressed, while its
lower edge will grow longer. What counteracts the tendency to sag
is chiefly the *tensile strength* (resistance to stretch) available along
the lower edge of the beam. Stone is not notable for tensile
strength, so the posts under simple stone beams must be closely
spaced. Aqueducts or bridges can be made in this way, but the
designer will soon discover that when the load they are meant to
bear is added to the weight of the transverse members, the posts
must be still more closely spaced. This wastes time and materials.

Roman engineers discovered a solution, the stone or masonry
arch (Diagram 1.2). Though domes had been built much earlier,
we shall see that the Roman arch provides the first analytic
approach to dome engineering. The arch is essentially a device for
dispensing with a center post, by splitting the thrusts a center post
would support and deflecting them to the sides. The downward
pull of gravity on the keystone is converted into paired outward
thrusts, which the face angles of successive stages transform into
downward thrusts once more, but downward thrusts now borne
by the side columns. Thus the columns actually support the
weight of the keystone and its neighbors, without having to be

1.1

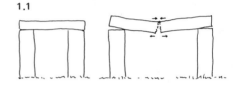

1.2

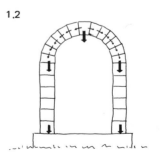

located directly under the stones whose weight they bear. So a central space is cleared beneath the arch.

It is clear that everything is held in place by weight, so that the continuities of stress are chiefly compressive.

Two or more intersecting arches will define a dome-shaped space, again clear of supporters because the work of support has been transferred to peripheral columns. The beehive-shaped tombs at Mycenae can be analyzed in this way. There the tendency of such arches to collapse outward is countered by, in effect, burying the dome and relying on the weight of tons of earth to sustain outward thrust. For a similar reason the stone dome of the Pantheon in Rome is enclosed in a huge masonry cylinder.

Though the visible continuities are compressive, there is in fact an invisible tension network which analysis cannot ignore. Each component of a stone dome is held in place by the earth's gravitational field, pulling tensionally "downward" through the structure. If the dome were inverted, the force that pulls it together would pull it apart. If it could be placed in orbit, it would drift apart. Thus its structural integrity depends on the weight of its components, and on the way they are oriented in earth's gravitational field. A successful design is essentially a feat of balancing. All forces are resolved along lines perpendicular to earth's surface, so that gravity and the mutual impenetrability of stones achieve a standoff. Any forces that deviate from this system of perpendicular resolutions will create a tendency to collapse inward or outward, and must be counteracted by braces or buttresses. Whether the placement of these is arrived at by rule of thumb, in the manner of the Gothic cathedral builders, or by sophisticated calculation in the manner of the twentieth-century engineer, their necessity says something about the precariousness of the structure's equilibrium, even when equilibrium is achieved without their aid.

If instead of discrete stones we use continuous curved beams of wood or metal for the arches, we arrive at the familiar ribbed domes of Saint Peter's in Rome or the Capitol in Washington, but we do not substantially alter the structural analysis. We greatly reduce the superincumbent weight, and we manage to separate the dome itself into "skin" (sheathing) and "bones" (trusswork), but we are still relying on compressive continuities to sustain most of the load. In certain respects, the efficiencies are less rational than in a stone dome: since the zenith of the arch no longer serves as a keystone, its chief function now is to load its supporters irrelevantly. The greatest concentration of structural members is at the zenith, where they have nothing to support, but instead constitute a problem for the members that support *them*. And successful design is still a feat of balancing. Unless thrusts are perpendicularly resolved, the dome will still tend to collapse inward or burst outward. Design usually elects to err in the latter direction, and the downward thrust at the zenith is translated into an outward thrust around the periphery (precisely where the structural members that ought to cope with it are most widely spaced). Here, in place of stone buttresses, a peripheral clamping ring holds things together. At Saint Peter's the system for coping with peripheral outward thrusts is reinforced by a huge iron chain which has kept the dome intact for four hundred years.

The Saint Peter's chain is a multi-tonned Band-Aid applied to a region of potential failure. A structure of almost any configuration can be designed on this principle: put it together somehow, and reinforce failure points as they appear. Failure points appear because portions of the structure impose an undue load on other portions: the load distribution is irregular and only accidentally related to stress-bearing capacity.

It is possible, however, to take a completely different approach. The way to do this is to abandon altogether the concept of structural weight impinging on the compressive continuity of bearing members, the whole guarded by occasional tensional reinforcement. Instead of thinking of *weight* and *support*, we may conceive the domical space enclosure as *a system of equilibrated omnidirectional stresses*. Such a structure will not be *supported*. It will be *pulled outward* into sphericity by inherent tensional forces which its geometry also serves to restrain. Gravitation will be largely irrelevant.

In a soap bubble or a balloon, an envelope of *surface tension* attempts to close inward against the outward compressive force of the enclosed air. The equilibrium between tension and compression is modeled as a spherical shape. In a hollow spherical structure, of which a dome is a section, the compressive forces, like the tensile, are incorporated into the skin itself, and their direction cannot be divided in so obvious a way between inward-tending and outward-tending. The tensile web supports the compressive members, and is also supported by it. The tensile pull can be as easily imagined tending outward as inward.

To understand this bootstrap effect, consider first a primitive tensile structure, consisting of two trees, a clothesline, and two poles (Diagram 1.3). The poles slant in opposite directions, and the system sketches a contained space.

1.3

Next, discard the trees, and fix the ends of the line to the earth, slanting the poles so that their lower ends and the anchor points of the line define a quadrilateral (Diagram 1.4). Provided the poles are prevented from slipping, this is perfectly stable, and we have framed a tent with no centerpole.

1.4

If we join the rope anchor points by a third pole, and replace the dotted lines on Diagram 1.4 with additional rope (Diagram 1.5), we shall find that we have a self-sufficient tension/compression system. The rope holds the poles both together and apart. The poles in turn lend shape to the prism-shaped rope network.

1.5

Here the reader should convince himself of the properties of this structure by experimenting with a simple model. Three dowels of convenient length (say, 9 inches) will do for the poles. Drive nails or pins into their ends and then tie them together as shown in Diagram 1.5, making the strings two-thirds the length of the dowels. As the last string is tightened, the tension network can be seen pulling the system outward into taut equilibrium. Thereafter the system resists deformation, and if deformed to an extent permitted by the elasticity of the tendons, will tend to restore itself to equilibrium.

The vertex points of this system, 6 in number, may be imagined as points on the surface of an enveloping sphere, since they are equidistant from a point in the center of the tensile prism. Additional members (poles and ropes, or struts and strings) can be so placed as to increase the degree of approximation to a sphere: we can make the system as spherical as we like. (*This will be discussed in detail later*.) As we do so, we shall find that the poles sketch the sphere's inner surface, the ropes its outer. In like manner, the stresses on the outer skin of a spherical structure tend to be tensile, and the stresses on its inner skin compressive. And the integrity of the spherical skin as a whole is wholly independent of central support. It is also independent of compressive load-bearing of the kind exemplified in post-and-beam construction or in the arch, since the compressive members are not in contact.

Now, return to the transition between Diagram 1.4 and Diagram 1.5 and note that structural integrity requires either a complete rope-and-pole system or else a partial system plus the earth. Tensional circuits must be completed somehow. Motion pictures of air-lifted geodesic domes show the bottom edge weaving and wavering until it is set on the ground and affixed there by fastenings.

The rope-and-pole prism shown in Diagram 1.5 is the simplest *Tensegrity* structure. (Tensegrity = *tens*ional int*egrity*.) It has no redundant components. All the domes described in this book, notably the numerous "geodesic" variants, exemplify special cases of Tensegrity principles. Their salient continuities are tensional, and their upper portions are not so much supported as *lifted* by tensional forces.

Unlike the stone arch or the stone dome, such structures are not made stronger by being made heavier. In fact, they can with advantage be made negligibly light in comparison with the tensional forces that bind the components. The one-way tension of terrestrial gravity is replaced by the multidirection tension of structural members. The system is therefore stable in any position.

Moreover, a tendency to peripheral or local stresses, such as those restrained by the chain round the dome of Saint Peter's, is supplanted by a multidirectional stress equilibrium. A corresponding multidirectional tension network encloses accidental stresses wherever they arise. There are no points of local weakness inherent in the system.

Tensegrity Prisms

We have noted that the structure developed in Diagram 1.5 is the simplest Tensegrity, consisting of 3 compressive struts and 9 tensile tendons (*tendon* = the portion of the tensile network between the two adjacent strut ends). It resembles a triangular prism one end of which has been rotated with respect to the other, thus twisting the quadrilateral sides. One additional strut (Diagram 1.6) will

1.6

convert the end triangles into squares; a further strut

1.7

(Diagram 1.7) will convert them into pentagons; and so forth. It is possible in this way to generate a potentially infinite family of T-prisms corresponding to the prisms of solid geometry.

We may imagine any such T-prism enclosed in a cylinder of height h and diameter d. The tendons (of length e) outlining the end n-gons are called *end tendons*. There are also n *side tendons*, of length t. (In general n denotes the number of struts, the number of side tendons, and the number of tendons bounding the end polygons.)

We shall assume that the end n-gons are equilateral. If they are equal to one another, the prism is *uniform*. If they

are unequal (though equilateral) the prism is *semiuniform*, and would be enclosed by a truncated cone instead of by a cylinder.

Though they join corresponding vertices of the top and bottom n-gons, the struts of any T-prism all lean uniformly, either clockwise or counterclockwise. That is because of the twist referred to above; the top polygon has been rotated with respect to the bottom polygon through an angle a called the *twist angle* (Diagram 1.8). Whatever the height or

1.8

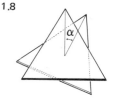

diameter of the structure, it can be shown that *for a given number of struts, the twist angle is constant* and is given by the formula

$$a = 90° - 180°/n. \qquad \text{[Eq. 1.1]}$$

This remarkable fact* makes it easy to calculate the lengths of struts and tendons for any values of n, h, and c.

One way to prove the twist-angle theorem is to use cylindrical coordinates. Diagram 1.9 shows the coordinate frame with 1 strut s, 1 side tendon t, and 1 end tendon e. Since the end tendon is one edge of the end n-gon, it subtends a center angle of $360°/n$. The cylindrical coordinates (r_1, ϕ, z) of A and B are thus $r_1, 0, 0$ and $r_1, 360°/n$, 0, respectively. Point C is not located above point B but is displaced counterclockwise by an additional angle a, the

*In effect demonstrated by Roger S. Tobie, "A Report on an Inquiry into the Existence, Formation and Representation of Tensile Structures" (Master's thesis, Pratt Institute: 1967).

1.9

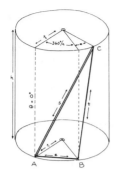

twist angle; its coordinates are $r_2, (360°/n) + a, h$. The distances AC and BC are strut length s and side tendon length t, respectively. The standard distance formula for cylindrical coordinates gives:

$$s = \sqrt{r_1^2 + r_2^2 - 2r_1 r_2 \cos\left(\frac{360°}{n} + a\right) + h^2}$$

and

$$t = \sqrt{r_1^2 + r_2^2 - 2r_1 r_2 \cos a + h^2}.$$

Rearranging the first equation, we obtain

$$h^2 = s^2 - r_1^2 - r_2^2 + 2r_1 r_2 \cos\left(\frac{360°}{n} + a\right),$$

and substituting this for h^2 in the second equation gives

$$t = \sqrt{s^2 + 2r_1 r_2 \left\{\cos\left(\frac{360°}{n} + a\right) - \cos a\right\}}.$$

To find out how this varies, we differentiate it, obtaining

$$\frac{\Delta t}{\Delta a} = \frac{r_1 r_2 \left\{\sin a - \sin\left(\frac{360°}{n} + a\right)\right\}}{\sqrt{s^2 + 2r_1 r_2 \left\{\cos\left(\frac{360°}{n} + a\right) - \cos a\right\}}}.$$

This is messy, but take heart. When the system is taut the tendon length t will be minimal, and the derivative above will = 0, signifying no rate of change. At this point the denominator drops out, since it must be the numerator that equals zero. And since the product of the radii cannot be zero, it can only be the functions of a that have vanished. So at the equilibrium state,

$$\sin\left(\frac{360°}{n} + a\right) = \sin a,$$

which is possible if

$$\left(\frac{360°}{n}\right) + a = 180° - a,$$

whence

$$2a = 180° - \left(\frac{360°}{n}\right)$$

and

$$a = 90° - 180°/n.$$

The twist angle therefore is a function of n alone, and is given by Equation 1.1, above.

If we now disregard such physical inconveniences as tendon stretch and deformations imposed by the weight of the structural members, we can use the cylindrical coordinate distance formula to calculate the dimensions of an idealized semiuniform T-prism. Putting r for $d/2$ — that is, one-half the diameter — we get:

$$\text{strut } (s) = \sqrt{r_1^2 + r_2^2 + 2r_1 r_2 \sin\left(\frac{180}{n}\right) + h^2}.$$

[Eq. 1.2]

$$\text{side tendon } (t) = \sqrt{r_1^2 + r_2^2 - 2r_1 r_2 \sin\left(\frac{180}{n}\right) + h^2}.$$

[Eq. 1.3]

$$\text{end tendon } (e) = d \sin(180°/n).$$ [Eq. 1.4]

If the T-prism is semiuniform there will of course be two end tendon lengths, corresponding to the two diameters.

EXAMPLE:
  Semi-uniform T-prism, 12 inches high, end diameters 6 inches and 10 inches: 5 struts. The equations give 14-inch struts, 12.7-inch side tendons, 3.5-inch tendons for the smaller end, 5.9-inch tendons for the larger end.

EXAMPLE:
  Uniform T-prism, 3 struts, and as high as it is broad. (Thus $h = d = 1$):
  End tendons are $\sin 60°$ or 0.86603; struts are 1.39,

side tendons are 1.03. Thus for a model 12 inches high, end tendons would be 10.4 inches long, struts 16.7 inches, side tendons 12.4 inches.

It may occur to us to want all tendons equal. In that case we put the right side of Equation 1.4 equal to the right side of Equation 1.3 and rearrange to give h: for $d = 1$,

$$h = \sqrt{\sin^2(180°/n) - \sin^2\left(45° - \frac{90°}{n}\right)}$$

[Eq. 1.5]

This proves to be solvable for $n = 1, 2, 3, 4, 5$. For $n = 6$ it yields 0, (that is, no height at all), and for $n > 6$ it yields a function of $\sqrt{-1}$, which is not modelable. Thus, while there is an infinite number of regular prisms, there are only three regular T-prisms, corresponding to $n = 3, 4, 5$. ($n = 1$ and $n = 2$ are structurally meaningless). Their proportions are:

1. $n = 3$.

      Strut/tendon ratio = 1/0.68125.

2. $n = 4$.

      Strut/tendon ratio = 1/0.6436.

3. $n = 5$.

      Strut/tendon ratio = 1/0.60843.

2. Spherical Tensegrities

The three-strut Tensegrity described in Chapter 1 is asymmetrical, having triangular ends and rhomboidal sides. It is the simplest member of an infinitely large family of Tensegrity prisms (T-prisms), analyzed in Appendix 1.1. All of these have rhomboidal sides and none can be made omnisymmetrical. Even the four-strut version (T-4 prism) has plane squares at two ends but twisted squares on four sides.

Since we are approaching the design of a dome, we want a Tensegrity whose face patterns, when projected onto a sphere, will divide the sphere's surface into symmetrically placed zones. The simplest Tensegrity of this kind employs 6 struts arranged in parallel pairs (Diagram 2.1).

2.1

Again the reader should stop and construct a model. There is no substitute for experimenting with an actual structure. Use dowels for the struts as before, and make the tendons 0.6 times their length. (For large models, the theoretical ratio is 1 : 0.6124.) When the 24 tendons are in place, it will be found that the parallel struts are spaced exactly half a strutlength apart, centerline to centerline. We also discover that when two parallel struts are moved toward or away from each other, the other two pairs move in exact concert. The entire structure expands or contracts symmetrically; it does not bulge here to accommodate a dimple there. The reason the struts will move at all is that the residual elasticity of the tendons is greatly multiplied by the geometry of the system. Elasticity Multiplication — to be examined later — is a characteristic inherent in Tensegrities.

This six-strut, twenty-four-tendon Tensegrity has no redundant parts. Each strut is held in place by the cooperative action of a system that comprises twenty-nine other members. When a strut is displaced by application of stress, the whole system undergoes symmetrical modification to accommodate the local movement. The system's symmetry is not deformed; the system *expands* as a whole or *contracts* as a whole. To permit this, each of the eight equilateral triangles rotates about its center; an inward or outward motion of the struts is accompanied by a rotary motion of the tension triangles. Obviously, this ability to respond *as a system* would be a valuable characteristic of large space-enclosure structures. Ability to respond as a system means that local stresses are being uniformly transmitted throughout the structure, and uniformly absorbed by every part of it. This principle points toward valuable economies in material. Instead of designing every part of the system to receive unassisted whatever loads it may incur, with consequent local accretions of weight and bulk, we may instead design the *system* on the assumption that local stress will be transmitted throughout its extent, and shared by all its members. The normal state of the system is not a state of rigidity but a state of equilibrium, to which, when disturbed, it seeks to restore itself.

Equilibrium

When the tension members of a Tensegrity are taut, it is in a state of equilibrium. To this state, however stressed, it always seeks to return. Unlike the "straightness" of Euclidean lines, the tautness of tension members is an approximation only. The weight of the cord or cable, however slight, will always curve it slightly; turnbuckles and fastenings have always residual slackness; materials have always residual elasticity. It is impossible to pull any line so tight that it could not, with sufficient effort, be pulled a little tighter. Hence the capacity of the system to absorb displacements and restore itself.

Still, for purposes of study and of system design, we can imagine an ideal state of equilibrium in which tendons pull perfectly straight and are perfectly tight, and predict the place every component would occupy if the ideal equilibrium could be realized with actual materials.

Examination of the six-strut Tensegrity discloses four tendons attached to each strut end. By recourse to a chemical metaphor, we may call it a *Valence-4 Tensegrity*; it is the simplest of a very large family of these. (The T-prism we examined in Chapter 1 has three tendons per strut end and is hence an example of the Valence-3 Tensegrity family.)

The design procedure for a Valence-4 Tensegrity commences from the fact that each strut lies at the bottom of a tensional "valley" whose sides are triangular. Other struts are attached to the apexes of these triangles (Diagram 2.2). Each triangle consists of two tendons plus the strut they share in common. Each pair of tendons, together with the strut, lies in a plane we may call a *tensile plane*. So *each strut lies along the intersection of two tensile planes*. (Two intersecting planes, according to projective geometry, suffice to determine the position of a line. This fact may help us understand why the struts tend to stay in place.)

We may now examine the conditions of system equilibrium (Diagram 2.3). Since equilibrium is not a static condition of being "at rest," but rather the resolution of forces that are pulling in several directions, the way to locate it is to examine the forces that are seeking to disrupt it. In the figure, which is Diagram 2.2 redrawn in a different perspective, S is the cross section of a strut lying at the bottom of a tensile valley ASB. Struts A and B are being pulled symmetrically outward, left and right, by other tendons in the system. If they were to obey this pull, they would lift strut S outward and upward. As this happened, angle ASB, the angle between the two tensile planes that locate strut S, *would tend to increase.*

But other tendons attached to strut S are restraining it from moving in this way. They are doing this by pulling it downward and inward. If they were to succeed, strut S, in moving downward, would pull struts A and B inward, and angle ASB, the angle between the two tensile planes that locate strut S, *would tend to decrease.*

So we can see that the forces in balance are determining angle ASB, the angle between the two tensile planes that locate strut S. Since the system is omnisymmetrical, the pairs of tensile planes associated with each strut will all be at this angle to one another.

2.2

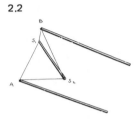

2.3

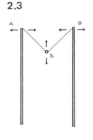

We may call it the *dip angle*, since the spherical contour of the system here dips inward, incising the polyhedron with a reentrant angle. There is a characteristic dip angle for every symmetrical Valence-4 Tensegrity system. We may intuit, and shall later prove in detail, that if we know the dip angle we can deduce all the system's component lengths and positions. So our first job is to find the dip angle of the system under consideration.

To deduce the dip angle of the six-strut Tensegrity system, we scrutinize the balance of forces more closely (Diagram 2.4). Concentrate on point A, and reflect that in the equilibrium state it stays where it is because all the forces that pull on it are in balance. We may specify some of these forces. The tendons in the tensile plane AS are pulling point A downward and rightward. The rightward and downward forces, represented by the heavy arrows, may be regarded as equal, like all the other sets of forces in the system. They are also at 90° to one another.

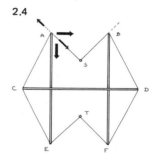

2.4

Point A does not move rightward because part of the effort of the tendons in the tensile plane AC is to pull it leftward. It does not move downward because part of the effort of the tendons in the tensile planes ET and CE is to pull strut AE upward. These pulls balance, and it stays where it is.

Now return to the rightward and downward forces represented by the heavy arrows at A. A point pulled upon by two equal forces will behave as though it responded to a single force, the angle of which is midway between them. So point A responds to a force pulling downward and rightward at 45° to the horizontal and vertical.

Similarly, the tensile plane BS pulls the point B downward and leftward at 45°. Being at 45° to the horizontal, the tensile planes AS and BS are at 90° to each other. The dip angle is the angle between these planes. Thus we have found that *in the symmetrical six-strut Tensegrity system the dip angle is 90°.*

We can now proceed more rapidly. Diagram 2.5 shows the same view of the system as Diagram 2.4, but with strut CD omitted and with dotted lines GH and JK drawn through S and T. GJ is the separation between struts S and T, and since the system is symmetrical this will be equal to GH, the separation between struts

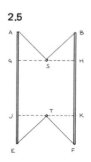

2.5

AE and *BF*. Moreover, since the dip angles *ASB* and *ETF* are both 90°, *AG = GS* and *JE = JT*. But we know from symmetry that *S* is the midpoint of *GH*, and *T* is the midpoint of *JK*. Since we have already found that *GH = GJ*, we see that *AG* and *JE* must each equal half of *GJ*. Adding *AG*, *GJ*, and *JE* (or 1/2 + 1 + 1/2), we learn that *GJ* equals half the length of the strut *AE*. But we know that *GJ* corresponds to the separation between the struts *S* and *T*. And all strut separations are alike. So we have shown that *in a symmetrical six-strut Tensegrity system the separation between any two parallel struts is exactly half the length of a strut*.

This fact defines the condition of the system when it is in equilibrium. The struts can only be displaced from half-length separation by a force from outside the system — an inquisitive hand, a falling tree, the pull of gravity on strut materials. Once displaced, half-length separation is the condition to which they will seek to return.

It is now a simple matter to calculate tendon lengths for taut-ness in the equilibrium condition. Diagram 2.6 shows how the pairs of struts are arranged in three mutually perpendicular (orthogonal) planes. In the six-strut Tensegrity system the ends of parallel struts need never be connected; a pair of struts joined by a tendon will always be at a 90° angle to one another. And by symmetry, all twenty-four tendons are of equal length. So there is only one tendon orientation to be considered, and we need only find the length of one tendon.

Diagram 2.7, a detail extracted from Diagram 2.6, shows the intersection *O* of the 3 planes, like the meeting of two walls and a floor. Half a strut *AE* is painted on one wall; the other half is below the floor. Half a strut *CB* is painted on the adjoining wall; the other half is in the far room. Thus *AE* and *CB* equal half a strut length each. *EO* is half the separation between two parallel struts, thus equals 1/4 strut length. Likewise *CO* = 1/4 strut length. Thus *AD = DC*. We now have the dimensions we need to solve the right-angled triangle *ADC*, the hypotenuse of which is a side of the right-angled triangle *ACB*. The hypotenuse of the latter in turn may be calculated; it is 0.6124[+]. This is the length of the tendon joining strut ends *A* and *B*.

2.6

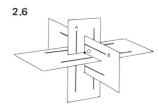

2.7

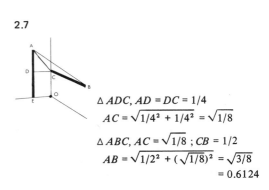

$\triangle ADC, AD = DC = 1/4$

$AC = \sqrt{1/4^2 + 1/4^2} = \sqrt{1/8}$

$\triangle ABC, AC = \sqrt{1/8} \; ; CB = 1/2$

$AB = \sqrt{1/2^2 + (\sqrt{1/8})^2} = \sqrt{3/8}$

$= 0.6124$

Thus *for strut length = 1, each of the 24 tendons of a symmetrical 6-strut Tensegrity system is* $\sqrt{3/8}$, *or 0.6124 approximately.*

We emphasize once more that this is an ideal Tensegrity system with weightless struts and ideally straight tendons. We have calculated design-center values, to which no actual model, large or small, will ever exactly correspond. In a large structure, say with 10-foot struts, the catenary sag of the tendons will always be effective though it may not be readily measurable. Moreover, strut weight will compress the system somewhat; triangles will rotate and tendons will stretch to accommodate this. And whether the model is made large, with beams and cables, or small, with dowels and string, perfect symmetry will prove impossible of attainment, however carefully dimensions are measured. These facts mean only that an actual Tensegrity structure is never quite in the calculated equilibrium state. We recall, however, from experimenting with a model that it has a unique ability to accommodate departures from equilibrium. It can be designed and constructed as though perfect equilibrium were attainable, and will simply accommodate to the deformities introduced by the physical characteristics of materials.

So we may want to know what we can predict about the system's behavior under stress.

Elasticity Multiplication

When the system is under stress and hence no longer in equilibrium, the pairs of parallel struts remain parallel but their separation is no longer 1/2 strut length. They have moved either closer together or further apart, and their separation is some new multiple of a strut length — call it s. As the struts move, the tendons stretch. Their length is no longer 0.6124 × strut length, but some greater multiple of a strut length: call it t.

To find how s and t are related, we may redraw Diagram 2.7 and label the distances CO and CD not 1/4 but $s/2$. (The length of the portions of struts we are dealing with is still 1/2; only the strut separation has changed, from 1/2 to s, so the half-separation shown in Diagram 2.8 is no longer 1/4 but $s/2$.) The right-angled

2.8

$AE = CB = 1/2$

$\triangle ADC,\ CD = s/2\ ;AD = 1/2 - s/2$

$\therefore AC = \sqrt{\dfrac{2s^2 - 2s + 1}{4}}$

$\triangle ACB,\ CB = 1/2, AC = \sqrt{\dfrac{2s^2 - 2s + 1}{4}}$

$\therefore AB = \sqrt{CB^2 + AC^2}$

$\quad = \sqrt{1/4 + \dfrac{2s^2 - 2s + 1}{4}}$

$\quad = \sqrt{(s^2 - s + 1)/2}$

triangle ADC now has $s/2$ and $(1/2 - s/2)$ for sides, and AC^2 becomes $(2s^2 - 2s + 1)/4$. Solving triangle ABC with the help of this, we find that AB, the tendon t, is $\sqrt{(s^2 - s + 1)/2}$, when s and t are multiples of strut lengths.

We know the value of t when the separation s is $1/2$ at the equilibrium condition. Sure enough, if we put 0.5 for s in this equation, we obtain $t = \sqrt{0.375}$, or 0.6123724357 (to ten places).

We have obtained a *general tendon-length equation* for the symmetrical 6-strut Tensegrity:

$$t = \sqrt{(s^2 - s + 1)/2}. \qquad \text{[Eq. 2.1]}$$

We are now ready to see what happens if we change the strut separation. Let us try increasing it by 1 percent, a modest increment, well within the stretching capacity of any likely tendon material. We may first define what we might expect to happen. Diagram 2.9 shows a pair of parallel struts 12 inches long. They are joined at both ends by strings 6 inches long, so that the strut separation is half a strut length, exactly as with a six-strut Tensegrity in the equilibrium condition. If we now force them apart an additional 1 percent, the separation will have increased from 0.5 to 0.505, a mere 0.06 inch. The strings will have no difficulty stretching that far, and we are unsurprised to find that the string length has also increased by the same percentage. To write this compactly, we may use the mathematician's usual symbol, Δ, for the amount of change. The change in strut separation, Δs, is 0.06 inch; the change in separation expressed as a percentage, $\Delta s\%$, is 1. And the tendon length changes are exactly the same: Δt is 0.06 inch, $\Delta t\%$ is 1. So $\Delta s / \Delta t = 1$, and $\Delta s\% / \Delta t\% = 1$.

To find out what happens when we move two parallel struts of a six-strut Tensegrity apart by the same amount, we have only to solve Equation 2.1, putting 0.505 in place of s. When we do this we find that t, the tendon length, has become 0.6123826418. The reader will now see why we calculated the equilibrium tendon length to ten places a page or so back: the difference between the tendon length at equilibrium (0.6123724357) and for strut-

2.9

spacing 1 percent greater than at equilibrium (0.6123826418) commences to show up only in the fifth place. The difference is 0.0000102061. In a model with 12-inch struts, separating the struts an additional 1 percent causes the strings to stretch about one ten-thousandth of an inch, which is quite unmeasurable. Dividing separation change by tendon stretch, we obtain not $\Delta s/\Delta t = 1$, but $\Delta s/\Delta t = 489.9$.

The elasticity of tensile materials is usually expressed as a percentage of a unit length. Converting the tendon stretch into a percentage, we find that each tendon has stretched 0.00166 percent while the strut separation was increasing 1 percent. So $\Delta s\%/\Delta t\% = 600$. If we handled the Tensegrity with our eyes shut, imagining that we were holding merely two sticks tied together, as in Diagram 2.9, we should think the tendon material was 600 times more elastic than it actually is. We are encountering the capacity of the system to *multiply the elasticity* of the tendon material.

We have here a classic case of *synergy*: behavior of whole systems, unpredicted by knowledge of the parts or of any subset of parts. Nothing we know about the struts and strings could allow us to predict their extraordinary behavior when they are united in a six-strut Tensegrity system.

It follows that calculations pertaining to such a system, taking into account the known characteristics of the materials but presuming the usual methods of assembling them, are certain to be wrong. We might suppose that if the struts were displaced by 10 percent a tendon would break, because our tendon material will not stretch 10 percent without breaking. But a little work with Equation 2.1 will show that a 10 percent strut displacement (from 0.5 to 0.55) changes tendon length by only 0.00102, a mere 0.167 percent. This is only 1/60 of what we might have expected ($\Delta s\%/\Delta t\% = 60$), and well within the elastic capabilities of any tension material we are likely to think of. By analogy, the tensile network hidden in geodesic domes quite defeats all normal calculations of their strength.

The reader will have noticed that when we increased the strut displacement from 1 percent to 10 percent, the ratio of separation change to tendon stretch ($\Delta s\%/\Delta t\%$) dropped from 600 to 60. This is generally true of Tensegrity structures: the elasticity multiplication is very great for small displacements (for instance 122,000 for a displacement of 0.002 percent), and drops rapidly as displacement increases (about 10 for displacement of 60 percent). Thus Tensegrities are extremely resilient under light loads. A complex Tensegrity model is never quite still, however tightly the tendons are stretched. On the other hand, it stiffens rapidly as loading increases.

We may also wonder what will happen if we push the struts of the 6-strut Tensegrity closer together instead of pulling them apart. The answer is, exactly the same thing. If we *decrease* the separation by 1 percent, to 0.495, and insert this value into Equation 2.1, we shall obtain the same tendon stretch as before, and the same $\Delta t\%/\Delta t\%$ ratio, 600. Whether we expand the system or contract it, the tendons stretch, and at exactly the same rate for the same percentage of change. So the system seeks equilibrium exactly as a ball seeks the bottom of a bowl. A graph of the s/t relationship is in fact bowl-shaped; more precisely, parabola-shaped. Any equation such as Equation 2.1, in which one quantity's variation is affected by the second power of another, will give us a parabola, if we plot points from it (Diagram 2.10).

The parabola has a theoretical zero point, where it touches the baseline, but this resembles the famous ideal point with position but no magnitude: we can never really say that it is occupied. In the real universe where winds blow, invisible forces tug, and molecules are in ceaseless motion, a Tensegrity will no more quite settle down than will anything else: there will always be minute motions, tiny strut displacements at the order of magnitude where elasticity multiplication is truly enormous and compensating forces have enormous advantage. The parabola's zero point is that ideal condition of rest which nothing real ever attains, and about which a Tensegrity in particular dances an eternal jig of pre-Socratic derision.

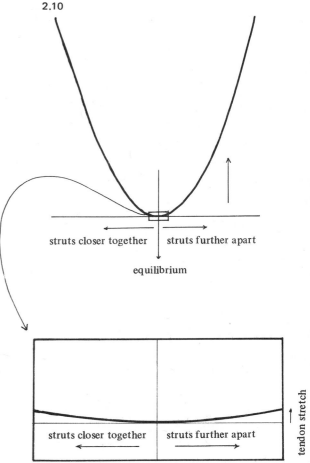

2.10

struts closer together | struts further apart

equilibrium

struts closer together ← | struts further apart →

tendon stretch

3. Complex Spherical Tensegrities

The simplest spherical Tensegrity, the six-strut Valence-4 version
with struts in three parallel pairs, has occupied us for many pages.
There are several reasons for understanding it in considerable
detail. (1) Tensegrity structures, and the geodesic domes of which
they supply first-approximation stress diagrams, have been
shunned by many investigators as unduly mystifying. It is worth-
while becoming convinced that their behavior yields to rational
analysis. (2) While dodges such as calculus would have shortened
the work, they can mislead unless we are sure we understand the
phenomena we are using them to describe. Simple mathematics
and experience of the directions in which strings pull taut will
suffice if we are patient. (3) The analysis has turned up some
principles that will help us with more complex structures. These
include:

- the property of behaving *as a system* in response to local
  events,
- the normal state as an ideal equilibrium about which
  observable behaviors oscillate,
- elasticity multiplication, a special case of synergy,
- the parabolic curve as a graph of system behavior,
- the value of the dip angle as a point of analytic attention.

Great Circle Tensegrities

Examining the symmetries of the six-strut Tensegrity, we may
think of surrounding it by three intersecting hoops, located like
the equator, the Greenwich meridian, and the 180° meridian
(Diagram 3.1). Each of these hoops — a great circle, or geodesic —

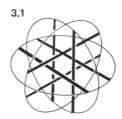

3.1

passes through both ends of one pair of parallel struts, and the three of them contain the system symmetrically. They resemble the great circles of the regular octahedron (Diagram 3.2), each of which contains four octahedron edges.

We may wonder if there are Valence-4 Tensegrity equivalents to other great-circle polyhedra. There are; but their number is severely restricted. Still, they have a great deal to teach us. We need polyhedra whose edges, projected onto a sphere, make great circles. We also need exactly four edges at each polyhedron vertex. There are just two polyhedra that will fulfill these conditions: the cuboctahedron and the icosidodecahedron (Diagram 3.3). Tensegrity equivalents to these polyhedra are shown in Diagram 3.4.

Once again, make models: preferably of both structures, but at least of the simpler (cuboctahedron). It is worth a little trouble. The Tensegrity cuboctahedron (abbreviated T-cuboctahedron) has twelve struts, arranged in four groups of three. Paint each group a different color before assembly. Use 1/4-inch or 3/16-inch dowel and make the struts 12 inches long. Drive pins into the dowel ends to anchor the tendons. Use thin, tough cordage: braided nylon fishline is excellent. (Avoid monofilament nylon, in which it is difficult to tie a nonslipping knot.) Diagram 3.5 shows a net diagram, and tendon lengths are indicated. Do as much as possible flat on the table; then join the indicated points, to close up the sphere.

Diagram 3.6 shows part of the net diagram for the T-icosidodecahedron. It uses thirty struts in six groups of five (six colors).

The T-cuboctahedron when completed should contain four equatorial triangles; the T-icosidodecahedron should contain six equatorial pentagons.

The T-cuboctahedron resembles the regular cuboctahedron in two principal ways. (1) Its struts describe four symmetrically placed great circles. (2) Its tendons outline the six squares and eight triangles of the cuboctahedron's faces.

We immediately notice two points of difference. (1) Whereas the cuboctahedron has six edges around each great circle, the T-cuboctahedron has only three struts. (2) The square and triangular faces outlined by the tendons do not abut on one another as

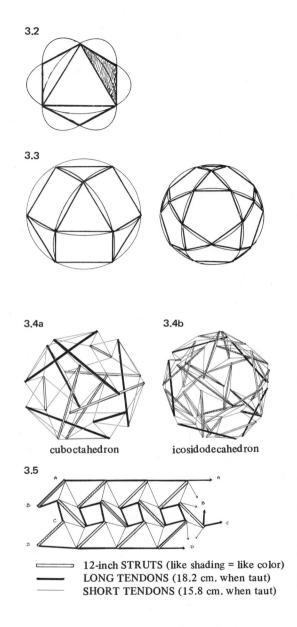

3.2

3.3

3.4a 3.4b

cuboctahedron   icosidodecahedron

3.5

| 12-inch STRUTS (like shading = like color) |
| LONG TENDONS (18.2 cm. when taut) |
| SHORT TENDONS (15.8 cm. when taut) |

3.6

Make co-linear struts same color (5 of each). Model requires 30 struts.

For 12-inch struts,
– Tendons outlining pentagons: 16.6 cm.
– Tendons outlining triangles: 15.3 cm.

Portion of T-Icosidodeca Net.

do the faces of the polyhedron. They are separated by diamond-shaped segments, each with a strut lying along its long axis and with two other struts facing each other end-to-end across its short axis. These differences from the parent polyhedron are general characteristics of Valence-4 Tensegrities.

Experimenting with the T-cuboctahedron, we may collect some observations about its behavior under stress. We remember a striking habit of the six-strut Tensegrity: when we grasped two struts that lay in the same plane and moved them together or apart, the other pairs of struts behaved in exact concert. Will this still work? Not quite. In the T-cuboctahedron two struts in the same plane will be adjacent sides of a circumferential strut triangle. If we grasp and wriggle two of these, action will occur throughout the system, but not uniformly. Watch the gaps between adjacent strut ends of the circumferential triangles. These will open and close as struts displace themselves. The gaps *within which* lie the two struts you are grasping will barely alter. Others, remote from the application of stress, open and close markedly. In general, disturbance is relayed to portions of the system far from where it entered. A ninety-strut 40-foot Tensegrity sphere erected at Princeton University in 1953 was struck by a snowplow at a point near the ground and exhibited component failure (a bent strut) high up on the other side, just 180° from the point of impact. In general, because stresses are diffused, component failure is rare. Light models of wood and string that look as though a cat could demolish them can be kicked, bounced, thrown, even accidentally stepped on, without damage.

If a model is stood on one of its triangular sides, another equivalent triangle will be uppermost. Lay a weight — a heavy book — on this upper face. As the model compresses it will rotate. The equatorial strut triangle neither bulges nor shrinks; it simply revolves. Others are deformed in complex ways. The net effect is compression of the sphere into an oblate spheroid, but *without increase in diameter*. The rotation of the equator takes up the forces that might be expected to expand it.

The *direction* of rotation under contractive stress is determined by the handedness of the system. Most Tensegrity systems exist as right- and left-handed mirror pairs; a particular model is either one or the other. It is easy to see (Diagram 3.7) that the three struts which rise from vertices of a triangular base can sketch either a clockwise or a counterclockwise rotation. Under compressive stress, these struts swing *away from* one another, either clockwise or counterclockwise according to the handedness of the system. The three struts angling downward from a triangular summit also swing away from each other, but since the upper cone is inverted with respect to the lower, their movement with respect to the equator is in the opposite direction from that of the base struts. Thus each equatorial strut is pulled both clockwise and counter-clockwise simultaneously; the stresses cancel, and the equatorial struts move neither out nor in. Thus the size of the equator (as we have noticed already) is unchanged. Instead the equator, and with it the whole system, is carried bodily around in the direction of rotation of the base struts. The only motionless zone is the base triangle, which is fixed to the floor, the tabletop, or the earth.

The six-strut Tensegrity of Chapter 2 is an exception to the general principle of right- and left-handedness. It has no handed-ness, hence no ring of cancellation, and so behaves with perfect symmetry throughout. All twelve of its vertices are displaced alike.

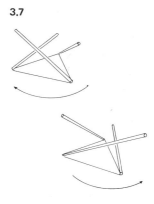

Dip-Angle Calculations

We next require a general procedure for calculating the dimensions and component lengths of spherical Tensegrities more complex than the six-strut one we have analyzed. The key to this, as before, is the dip angle.

Examining the six-strut structure in Chapter 2, we discovered that the dip angle between parallel struts was necessarily 90°. Diagram 3.8 shows how to extend this reasoning. It depends on something we can readily observe by inspecting a model, that any strut we may concentrate on is staying where it is because two opposed forces are balanced out. The two tensile planes at whose intersection it lies are bounded by four tendons whose net effort is to pull it straight outward, away from the center of the system.

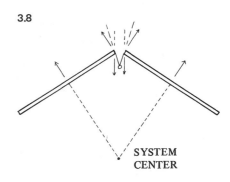

SYSTEM
CENTER

It is anchored at its ends, however, by tendons whose net effect is to pull it straight inward, toward the center of the system. These forces can be diagrammed right at the end of the strut, where all tendons are attached; Diagram 3.8 represents them by light arrows. The tendons in the tensile plane align themselves midway between these forces.

Diagram 3.9 shows how the situation would look if the struts were extended until they touched. They enclose an angle we may call θ. The rest is simple geometrical reasoning, shown with the diagram. We conclude that whatever the angle, θ, at which two struts meet, the dip angle δ will be

$$\delta = 90° - \theta/2. \qquad \text{[Eq. 3.1]}$$

These two struts are sides of a regular polygon which lies in the plane of a great circle round the models we have examined, but which may lie in a lesser circle plane round Tensegrities we shall encounter later. In either case, the angle θ between two adjacent struts is the interior angle between any two sides of a regular polygon with n sides. This is $180° - (360/n)$. Equation 3.1 tells us that the dip angle δ is $90°$ minus half this, or $90° - 1/2[180° - (360°/n)]$. Simplified, this gives us a compact alternative form for the dip-angle equation:

$$\delta = 180°/n. \qquad \text{[Eq. 3.2]}$$

So we have only to divide $180°$ by the number of struts in one plane of a Tensegrity sphere (otherwise put, the number of struts round a great circle or lesser circle) to obtain the dip angle between any two of these struts.

The six-strut Tensegrity of Chapter 2 has two struts in each plane. The dip angle between them is thus $180°/2$, or $90°$, the same result we got earlier by special-case reasoning.

In the twelve-strut T-cuboctahedron, three struts (an equilateral triangle) lie in any plane. Thus the dip angle is $180°/3$, or $60°$.

In the thirty-strut T-icosidodecahedron, five struts (a pentagon) lie in any plane. The dip angle is $180°/5$, or $36°$.

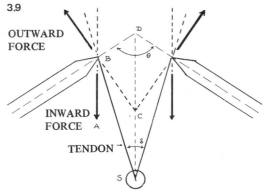

3.9

OUTWARD FORCE

INWARD FORCE

TENDON

$\angle ABC = \angle BCD = 90° - \angle BDC$
$= 90° - \theta/2$
$\angle BSD = 1/2 \angle ABC$
$= 1/2 (90° - \theta/2)$
$\therefore \delta = 2 \angle BSD = 90° - \theta/2$

Calculation of Remaining Dimensions

We shall now see how the dip angle can be used to calculate all the remaining dimensions of a Valence-4 Tensegrity sphere.

It is convenient to call the strut length 1 and express all other dimensions in strut lengths. If calculation shows that a certain tendon length is 0.523, this means it is 0.523 × strut length. If we are using 8-foot struts, this tendon will be 8 × 0.523, or 4.184 feet or 50.2 inches.

For model-building, we normally start with known strut lengths, anyhow. For larger structures, we are more likely to start with a desired diameter. No problem. Equation 3.5, below, relates radius to strut length and so gives us access to all the other dimensions, even when the size of the structure we want to build and its geometry are the only two things we know.

Proofs of all equations in this chapter are given in the chapter appendix. Elementary geometry and trigonometry suffice to yield all of them.

(1) We obtain dip angle δ

$$\delta = 180°/n, \qquad \text{[Eq. 3.2]}$$

where n = number of coplanar struts around the structure.

(2) We derive the gap g. The gap (Diagram 3.10) is the linear distance between the ends of two adjacent struts on a greater or lesser circle.

$$g = \sin^2 (\delta/2). \qquad \text{[Eq. 3.3]}$$

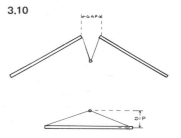

3.10

(3) We next find the dip. The dip is the median of a tensile plane: the distance (Diagram 3.10) from a strut end to the midpoint of the strut that crosses beneath it. Its symbol is d, and

$$d = \sin/2 (\delta/2). \qquad \text{[Eq. 3.4]}$$

We pause to notice that tendons which simply follow the dip, joining the ends of adjacent struts to the midpoint of the strut passing below, will suffice to make a stable structure with no other tendons whatever. In these center-connected Tensegrities, *whose only tension members are little V-shaped slings, the tension network is no longer optically continuous but must be traced in*

*part through the struts. This is not a recommended method of
construction because it imposes a bending load on the struts. It is
a good way to analyze certain features of Tensegrity construction.
It particular, it draws our attention to the fact that the dip is the
crucial parameter. When the dip slings in such a model are correctly dimensioned, the gaps and strut angles will automatically adjust
themselves into the equilibrium state, and all struts around any
great circle or lesser circle will align themselves in a common plane.*

(4) We now obtain the radius, r. This means the radius of the
circumsphere: a sphere into which the Tensegrity would exactly
fit, all its strut ends touching the spherical envelope. Or it is the
distance outward from the center of the system to the end of any
strut. The equation is brief:

$$r = \sqrt{(1 + 3g)/16g} \qquad \text{[Eq. 3.5]}$$

This gives us the radius expressed in strut lengths. Should we wish
to design a sphere of given radius R, then the strut length is, of
course, R/r.

(5) Our next step is to obtain an arbitrary quantity P. Its whole
purpose is to simplify the tendon-length equations by working
out separately a portion common to each of them.

$$P = \sqrt{(g - g^2 + 1)/4} \qquad \text{[Eq. 3.6]}$$

(6) Diagram 3.11 shows why we frequently need two different
tendon lengths. When strut circles cross one another at 90° angles,
tendon lengths are equal. When they cross at some other angle,
long and short tendons alternate around each strut. To find the
two lengths, we shall need the angle ι at which the planes bounded
by strut circles cut one another: more precisely, we need the
cosine of this angle. The axes of great-circle and lesser-circle planes
are a polyhedron's axes of symmetry, and since all regular and
semiregular polyhedra have either oc-tet or icosahedral symmetry,
there are only three possible values for cos ι:

- Circle axes at 90°. Cos ι = cos 90 = 0.
- Octahedral symmetry, circle axes at 70.5288°. Cos ι = 1/3.
- Icosahedral symmetry, circle axes at 63.43495°. Cos ι = 0.4472.

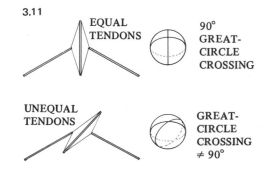

3.11

EQUAL
TENDONS

90°
GREAT-
CIRCLE
CROSSING

UNEQUAL
TENDONS

GREAT-
CIRCLE
CROSSING
≠ 90°

Diagram 3.12 shows the simple case of the cuboctahedron. It is octahedrally symmetrical, and we may expect cos ι to be 1/3. That is because the angle at which great-circle planes intersect is the dihedral angle of a tetrahedron, or 70.5288°, the cosine of which is 1/3. Tendons calculated by using this value in Equation 3.7 and Equation 3.8, below, will be taut and the model symmetrical, with circumferential struts aligned in common planes.

We pause to note one complication, more likely to affect the model-builder than the dome engineer: the essentially static mathematics of polyhedra will not always coincide with the dynamics of a Tensegrity. This is apt to show up whenever we encounter edges that lack a triangular face on at least one side. For instance, the small rhombicosidodecahedron, a Valence-4 polyhedron, consists wholly of intersecting lesser circles which cut it into square, pentagonal, and triangular faces. It yields a handsome Tensegrity sphere of sixty struts. Since the twelve lesser circles are icosahedrally coordinated, cos ι is 0.4472, and Equation 3.7 and Equation 3.8 will yield long and short tendon lengths. If we construct the model, however, we shall find (Diagram 3.13) that each strut has a triangular face along one side of only one end. The triangle pulls the associated square face into a diamond shape in such a way as to make all strut intersections equiangular. To make the model symmetrical we must refigure the tendons as though cos ι were 0, and all tendons of equal length. Methods of foreseeing such effects are of such limited application as not to seem worth the space required to expound them. *Check all Tensegrity predictions by building models.*

(7) We may now figure the long and short tendons, t_1 and t_2.

$$t_1 = \sqrt{P^2 + (g/2)^2 + (Pg \cos \iota)} \qquad \text{[Eq. 3.7]}$$

$$t_2 = \sqrt{P^2 + (g/2)^2 - (Pg \cos \iota)} \qquad \text{[Eq. 3.8]}$$

When the great-circle axes cross at 90° (cos ι = 0), we have a single tendon length t, obtained very simply:

$$t = \sqrt{(g + 1)/4}. \qquad \text{[Eq. 3.9]}$$

To illustrate this latter situation, consider the six-strut Tenseg-

ANGLE BETWEEN GREAT-CIRCLE PLANES EQUALS THE DIHEDRAL ANGLE OF A TETRAHEDRON = 70.5288°

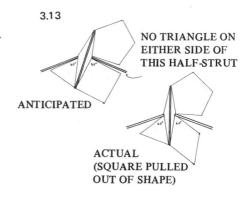

NO TRIANGLE ON EITHER SIDE OF THIS HALF-STRUT

ANTICIPATED

ACTUAL (SQUARE PULLED OUT OF SHAPE)

rity of Chapter 2. It has two struts in each plane, so by Equation 3.2 the dip angle is 180/2, or 90°. The gap, by Equation 3.3, is $\sin^2 (90°/2)$, or 0.50. This is another way of stating what we learned before, that the separation between parallel struts is half a strut length. And the tendon length, by Equation 3.9, is $\sqrt{(0.5 + 1)/4}$, or 0.6124, which checks with what we obtained by geometrical reasoning.

Table 3.1 illustrates the other two symmetry cases, deriving step by step the dimensions given in Diagram 3.5 and Diagram 3.6 for the T-cuboctahedron and the T-icosidodecahedron.

A sophisticated pocket calculator makes such computations so simple it is easy to become addicted to their multidecimal precision. Pause to reflect what it is we are calculating. We are describing to a high degree of accuracy the *equilibrium state* of an ideal Tensegrity sphere *built of weightless materials*. In the equilibrium state so envisaged, the tendon network pulls only against the ideal incompressibility of the compressive struts. In a real structure the one-directional tensile field of earth's gravity pulls through the multidirectional tensile field of the structure, imposing flattenings and compensatory rotations, as described earlier in this chapter. Tendons, however tightened, will describe catenaries, not straight lines, and two tendons will not meet at a point but at a fastening of measurable size.

Moreover, no one but a model-builder or a sculptor is likely to construct a complete sphere. Tensegrity domes are quite practical, and inspection of spherical models will suggest the simplicity of equatorial truncation. Groundlevel anchor points for certain tendons, and the thrust angles of half-struts embedded in the ground, are easily determined. Since with the aid of a few strategically placed turnbuckles it is relatively easy to lengthen or shorten tendons in the field, the designer may safely calculate dimensions from the equations, compensating where necessary for modes of fastening, and make adjustments on the actual structure.

TABLE 3.1

	T-Cuboctahedron	T-Icosidodecahedron
	3 struts per great circle	5 struts per great circle
Dip angle δ	$180°/3 = 60°$	$180°/5 = 36°$
Gap g	$\sin^2(60°/2) = 0.25$	$\sin^2(36°/2) = 0.09549$
Dip d	$\sin/2(60°/2) = 0.25$	$\sin/2(36°/2) = 0.1545$
Radius r	$\sqrt{(1+3[0.25])/16[0.25]} = \sqrt{0.4375} = 0.6614$	$\sqrt{(1+3[0.09549])/16[0.09549]} = \sqrt{0.842} = 0.9176$
P	$\sqrt{(0.25-0.25^2+1)/4} = \sqrt{0.2969} = 0.54486$	$\sqrt{(0.09549-0.09549^2+1)/4} = \sqrt{0.27159} = 0.52114$
$\cos\iota$	$\cos(2\text{ arc sin }[1/(2\sin 60°)])$ $= \cos 70.5288° = 1/3$	$\cos(2\text{ arc sin }[1/(1+\sqrt{5})\sin 36°])$ $= \cos 63.43495° = 0.44721$
Long tendon t_1	$\sqrt{0.54486^2 + (0.25/2)^2 + (0.54486)(0.25)(1/3)}$ $= \sqrt{0.3579} = 0.5983$	$\sqrt{0.5211^2 + (0.09549/2)^2 + (0.5211)(0.09549)(0.44721)}$ $\sqrt{0.29607} = 0.5442$
Short tendon t_2	$\sqrt{0.54486^2 + (0.25/2)^2 - (0.54486)(0.25)(1/3)}$ $\sqrt{0.2671} = 0.5168$	$\sqrt{0.5211^2 + (0.09549/2)^2 - (0.5211)(0.09549)(0.44721)}$ $\sqrt{0.2516} = 0.5016$

Thus, a model using 12-inch struts would require tendon lengths of 18.2 cm. and 15.7 cm. and would have a radius of 7.9 inches.*

Thus, a model using 12-inch struts would require tendon lengths of 16.6 cm. and 15.3 cm. and would have a radius of 11.3 inches.*

A 10-foot-diameter structure (radius 5 feet) would have struts 7 1/2 feet long and use tendons 54 inches and 46.8 inches long.

A 20-foot-diameter structure (radius 10 feet) would have struts 10.89 feet long and use tendons 71 inches and 65.6 inches long.

*Dowel comes in 3-foot lengths, but tendons are easier to measure in centimeters.

Derivation of Tensegrity-Sphere Equations

Why the dip angle assumes the value it does, we have discussed in the text. This appendix shows in detail why the gap, dip, radius, and tendon lengths necessarily follow from the dip angle. The reader who is content simply to use the equations is at liberty to skip these proofs.

Diagram 3.14 shows two struts, with tendons at their ends making a dip angle δ, and radii from their centers to the center O of the system, where they make the center angle α. Since all the center angles in the system will total $360°$, for n struts $\alpha = 360°/n$. But from Equation 3.2, $\delta = 180°/n$, thus $\delta = \alpha/2$.

3.14

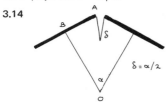

We move on to Diagram 3.15, which shows half a strut AB, two radii BO and CO with the angle $\alpha/2$ between them, a half dip angle, ACD, and a half gap, AX. Angle AXC and Angle ABO are both $90°$, and AC is the dip.

We have shown that $\delta = \alpha/2$; hence Angle $BOC = \delta$, and Angle $ACD = \delta/2$. Since $BO = CO$, triangle BOC is isosceles, and Angle $BCO = (180° - \delta)/2$. But since Angle $ACD = \delta/2$, Angle ACB, which is a straight line minus both Angle ACD and Angle BCO, must be $90°$. Thus ABC is a right-angled triangle, and $\sin ABC = AC/AB = $ dip/half-strut. Calling a strut length 1, dip $= 1/2 \sin ABC$. But since Angle $ABO = 90°$, and Angle $CBO = $ Angle $BCO = (180° - \delta)/2$, Angle $ABC = 90° - [(180° - \delta)/2] = \delta/2$. Thus:

$$\text{dip} = \sin/2 \, (\delta/2). \qquad [\text{Eq. 3.4}]$$

The gap equation follows readily. AX is half the gap, and

Angle AXC is $90°$. So in the right-angled triangle AXC, where the angle at C is $\delta/2$, $\sin(\delta/2) = (\text{gap}/2)/\text{dip}$. Thus the gap $= 2 \times \text{dip} \times \sin(\delta/2)$, or $2\,[\sin/2\,(\delta/2)]\,[\sin(\delta/2)]$, which simplifies to

$$\text{gap} = \sin^2(\delta/2). \qquad [\text{Eq. 3.3}]$$

3.15

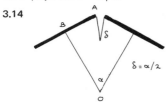

$BO = CO = $ radius
$AC = $ dip
$AX = 1/2$ gap

To derive the radius, we redraw the previous diagram (as Diagram 3.16) inserting (AO) the circumsphere radius we wish to find. It is the distance from the center O to the tip of a strut. Various things we know already are written beside the diagram. We see immediately that in the right-angled triangle ABC, Side $BC = \sqrt{1/4 - d^2}$. We also see that in the isosceles triangle BOC, Side $BO = BC/2 \times \sin \delta/2$. Thus $BO = \left(\sqrt{1/4 - d^2}/2\right)\sin(\delta/2)$. Finally, in right-angled Triangle ABO, the hypotenuse $AO = \sqrt{1/4 + BO^2}$. Substituting into this the value we have just obtained for BO, we get $r = \sqrt{[1 - 4d^2 + 4\sin^2(\delta/2)]/16\sin^2(\delta/2)}$. Since $d^2 = g/4$ and $\sin^2(\delta/2) = g$, this reduces to

3.16

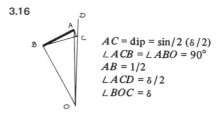

$AC = \text{dip} = \sin/2\,(\delta/2)$
$\angle ACB = \angle ABO = 90°$
$AB = 1/2$
$\angle ACD = \delta/2$
$\angle BOC = \delta$

$$\text{radius} = \sqrt{(1 + 3g)/16g}. \quad \text{[Eq. 3.5]}$$

Diagram 3.17 and Diagram 3.18 show how the equations for the tendon lengths are derived. Diagram 3.17 is a perspective drawing showing half a strut, the gap, the associated dips, and the two tendons. We have also drawn a perpendicular, *BA*, from the strut to the midpoint of the gap (thus *EA* = *AC*) and joined *A* to *D*.

The working is now fairly obvious. In right-angled Triangle *ACB*, we know *AC* and *CB* and can calculate *AB*. In the right-angled Triangle *ABD*, we know that *DB* = 1/2,

and having derived *AB* can calculate *AD*. This is the quantity *P* of Equation 3.6.

We then turn to Diagram 3.18, in which *EA* and *AC* are each $g/2$ and *AD* = *P*. In Triangle *EAD* we thus know two sides and the cosine of the included angle, and the standard formula $ED^2 = EA^2 + AD^2 - 2(EA)(AD)(\cos \iota)$ yields us Equation 3.8 for the short tendon *ED*. The equation for the long tendon *CD* is then derived from the fact that $\cos \iota = -\cos(180° - \iota)$.

3.17

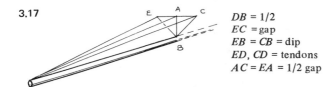

DB = 1/2
EC = gap
EB = *CB* = dip
ED, *CD* = tendons
AC = *EA* = 1/2 gap

3.18

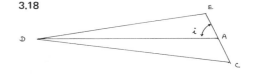

4. Tendon System Minima

Commencing with a clothesline and two poles, we have considered in turn elementary Tensegrity systems (Chapter 1), the six-strut spherical Tensegrity (Chapter 2), and twelve-strut and thirty-strut spherical Tensegrities (Chapter 3), the configurations growing ever more suitable for slicing off to yield practical domes. We are gradually deriving the structural principles of the geodesic dome from the Tensegrity sphere, of which it is a special case. The reader will reap the benefits of this approach when he arrives at the geodesic dome prepared to think of its sustaining structural continuities instead of assuming that rigid framing members exist to bear the weight of other rigid framing members, like studs in a frame house or bricks in a wall.

In two more steps we shall have arrived at domes in which things are bolted together instead of slung from cables. The reader anxious to get there may skip the present chapter, which refines our understanding of Tensegrities by describing in another way the forces they contain. The reader who finds Tensegrity independently fascinating is invited to tarry.

Six-strut Tendon Minima

In Chapter 2 we derived the equilibrium condition of the six-strut Tensegrity by concentrating on the sets of opposed forces whose compromise causes the dip angle to settle where it does. In Chapter 3, by a generalized version of the same reasoning, we arrived at a simple equation for predicting the dip angle of a multistrut Tensegrity sphere. We also learned that the dip angle governs all the other dimensions of the Valence-4 Tensegrity systems, so that knowledge of the dip angle enables us to calculate them.

It might have occurred to us to approach the initial problem in

Chapter 2 differently. We might have reflected that when we build a six-strut system we pull all the tendons tight, so tight that they cannot be pulled any tighter. It appears, then, that when the Tensegrity is in equilibrium *the total length of the tendon system is minimal.* This deduction is reinforced by our discovery that however we disturb the system we stretch the tendons; hence in equilibrium they must have been as short as possible. If we could arrive at the conditions for minimum tendon length, we should have another way of calculating the critical dimensions of the system.

We have seen what governs change in tendon length: it is displacement of struts. We can imagine the three pairs of parallel struts all arrayed in space at some large separation, and then commencing to approach one another symmetrically. The ends the tendons will join also approach one another. Eventually the centers of the struts will cluster at the center of the system. But some time before that happens, the ends the tendons will join will have come as close together as they are going to and will have started to move apart once more. It is that moment of maximum proximity (minimum separation) that we want to fix.

When we came upon elasticity multiplication in Chapter 2, we discovered (Equation 2.1) that whatever the separation s between parallel struts, the distance each tendon will span is $\sqrt{(s^2 - s + 1)/2}$. This equation is not linear (Diagram 2.10 shows that it is represented by a parabola). The rate at which the tendon length changes, levels off slowly until for a critical instant there is no change at all (the zero point of Diagram 2.10; the tendon minimum we are pursuing). Then the rate of change increases once more, slowly at first, then faster. Anywhere along the parabola in Diagram 2.10, a steep slope designates rapid change, a shallow slope slow change. At the elusive instant of horizontality (zero slope) the rate of change is zero.

None of this is news to the freshman calculus student, who learned on the first day of class that the rate of the change going on in an equation is described by the *derivative* of the equation, and a few days later was able to write the derivate of $s^2 - s + 1$. It is $2s - 1$. When the quantity is minimal it has momentarily ceased to change: its rate of change (hence its derivative) is zero. This will happen when $2s - 1 = 0$, that is, when $s = 1/2$.

This means that *when the distance between parallel struts in a symmetrical six-strut Tensegrity is 1/2, the distances to be spanned by the tendons, and hence the lengths of all tendons, are minimal.** This half-strut-length strut separation agrees with what we learned in Chapter 2 by a different chain of reasoning.

To learn the length of this minimum tendon, we have only to return to Equation 2.1 and write $1/2$ for s.

$$t = \sqrt{(1/2^2 - 1/2 + 1)/2}.$$

Then $t = \sqrt{0.375} = 0.6124$, which also agrees with the result we obtained in Chapter 2.

So *for a symmetrical six-strut Tensegrity system the assumption that the dip angle will be 90° and the assumption that the tendon system will be of minimum length yield exactly the same results.*

It is not difficult to show that this principle holds true of complex spherical Tensegrities as well. Consider Diagram 4.1, which shows half a strut BC with its midpoint C at a distance OC from the center O of the system. AB is the dip, which terminates at the midpoint of another strut shown in cross section at A. Since all strut midpoints are equidistant from the center, $OC = OA$. Angle θ, bounded by OD, OE, is obviously $360°/2n$, where n is the number of struts round a circle. The angle α is half the angle subtended by a strut at the center of the system and will increase as the strut is pulled in toward the center. As this happens, point A will also move in, since all struts move uniformly, and the dip AB will change. Hence the angle α tells us how far each strut is from system center and gives us an alternative way of defining the equilibrium state.

We can define the dip as a function of angle α:

$$\text{Dip} = AB = \left(\sqrt{2 - 2\cos\theta - 2\sin\theta \tan\alpha + \tan^2\alpha}\right)/2\tan\alpha.$$

[Eq. 4.1]

4.1

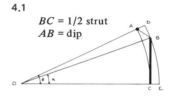

BC = 1/2 strut
AB = dip

*This simple approach was pointed out to me by Professor Rufus Isaacs.

Differentiated with respect to α, this gives

$$\frac{\Delta AB}{\Delta \alpha} = \frac{\csc^2 \alpha \, (-2 + 2 \cos \theta + \sin \theta \tan \alpha)}{2 \sqrt{2 - 2 \cos \theta - 2 \sin \theta \tan \alpha + \tan^2 \alpha}}.$$

[Eq. 4.2]

Setting this equal to zero and discarding two irrelevant possibilities, we find that for minimum tendon length,

$$\alpha = \arctan(2 \tan[\theta/2]) = \arctan(2 \tan[90°/n]).$$

[Eq. 4.3]

If we now substitute this for α in Equation 4.1, we discover after some tedious routine work that when the tendon length is minimal, the dip is $1/2 \sin(90°/n)$, which a glance at Equation 3.4 will show us is exactly what we arrived at in Chapter 3 by the dip-angle method.* And all other dimensions follow from that of the dip.

So *when any complex spherical Tensegrity is at its equilibrium state, all tendons, and hence the total tendon system, will be of minimum length.*

This tells us rigorously what we may have intuited, that we bring a Tensegrity to its equilibrium state by pulling everything as tight as possible. Thereafter any outward or inward force, in attempting to make the system larger or smaller, must also strive to make the tendons longer and will be inhibited by their restoring elasticity. Elasticity multiplication will permit the system to yield more than we might expect, but it will always seek to restore itself to equilibrium.

*I am indebted to Roland O. Davis of Goleta, California, for these calculations.

5. Geodesic Subdivision

We have been increasing the number of struts in a spherical Tensegrity system, from six to twelve to thirty. The systems have been growing more spherical, because they have more vertices, closer together, and the chords connecting these vertices (struts and tendons) are less and less different from arcs. In addition to making the spherical surface smoother, and hence easier to equip with a weather-break, the progressive shortening of structural members tends to make large structures increasingly practical.

Still, we seem to have reached a limit. The inventory of uniform polyhedra which are both bounded by great circles and equipped with four edges per vertex is exceedingly short. In fact, we have now employed the only three: the octahedron, the cuboctahedron, the icosidodecahedron, corresponding respectively to the six-strut, twelve-strut, and thirty-strut Tensegrity spheres. We might explore existing *lesser*-circle systems; indeed we have briefly discussed the small rhombicosidodecahedron, which will make a sixty-strut Tensegrity sphere. If we make this model, or even diagram it, one disadvantage appears immediately: it has no appealing natural lines of truncation. Slicing it on the plane of a lesser circle, we acquire two relatively useless pieces: an extremely shallow saucer, and a 7/8 sphere like an inverted goldfish bowl. Another disadvantage is longer-range: we have no reason to expect the inventory of Valence-4 lesser-circle polyhedra to last much longer than the great-circle series did.

If we are not to be constantly frustrated by the nonexistence of the sort of polyhedra we want, we shall need a way of devising new ones to our specifications.

This possibility does not coilide with the geometer's contention that polyhedra come in small, finite sets. The sets are small because the geometer stipulates that faces of the same shape must be exactly alike. This imperative is rather aesthetic than structural, and once we relax it, we shall find at our disposal literally infinite arrays of *nearly* uniform polyhedra, as multiedged and as elaborately faceted as we may wish. We acquire them by systematically subdividing the faces of existing polyhedra.

Diagram 5.1 shows a regular octahedron, having eight equilaterally triangular faces. Diagram 5.2 shows each face of the octahedron divided into four smaller triangles, by simply connecting the midpoints of the edges. These thirty-two small triangles are also equilateral, but the figure does not look promising as a way of generating anything spherical.

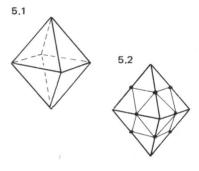

5.1

5.2

The reason, of course, is that the vertices and midedge points of the octahedron are at different distances from its center. If we call the center-to-vertex radius 1, the center-to-midedge radius is only $\sqrt{2}/2$, or 0.7071.

But there is no reason at all why the twelve vertices that lie on the octahedron's midedges cannot be pushed out until they are as far from the center as the eight vertices that coincide with the octahedron's vertices. The radii to all twenty vertices will then be identical (= 1), as if the vertices all lay on the surface of a sphere, and the edges that connect them will be chords of that imaginary sphere.

Diagram 5.3 shows what this would look like. Notice that we are not envisaging spherical triangles. The sphere implied by the drawing is nothing but a handy way of reminding ourselves that all vertex points are at a uniform distance (called 1) from the center of the system. The lines that bound the triangles are straight, not curved; chords, not arcs.

5.3

We can concentrate now on one octahedron face. Diagram 5.4 shows what has happened to it. The midpoints M, M, M of the sides have been pushed out until their distance OM_1 from the center O is the same as the distance of A, B, and C from the center O. It is clear that the triangle M_1, M_2, M_3 is equilateral, since it still corresponds in shape to the equilateral MMM. In moving out, however, its edges have grown larger, by the same geometry which gives a larger image when the screen is farther from the projector. But AM_2 has not grown by so much, since its upper end is still

5.4

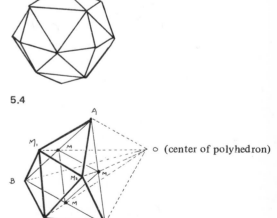

o (center of polyhedron)

anchored at A; it is like an image projected on a tilted screen. The same is true of AM_1. So the top triangle, AM_1M_2, is not equilateral but isosceles.

If we want data on this polyhedron, we thus have two edge lengths to calculate. Diagram 5.5 shows a way to go about it. Two applications of Pythagoras's theorem will give us those isosceles legs, and the knowledge that similar triangles are similar in all respects will give us the sides of the equilateral central triangle.

So we have devised a polyhedron with $8 \times 4 = 32$ triangular faces. Eight of them are equilateral, with sides equal to the polyhedron's radius ($= 1$). Twenty-four are isosceles, with bases equal to 1 and two legs equal to 0.7654^+. We obtained this by dividing each edge of the octahedron in two. We call this process *a two-frequency division of the octahedron*, and we call the resulting polyhedron *a two-frequency octahedron* (2ν octa for short).

There is no theoretical limit to this process of frequency subdivision. Diagram 5.6 shows how to generate one face of a four-frequency octahedron (4ν octa) and Diagram 5.7 shows the polyhedron we obtain. Diagram 5.8 lists edge lengths for the sixteen triangles we derive from each octa face. (Ways of calculating them will occupy us later.) They are of four different shapes: one equilateral (*EEE*) at the center; three isosceles (*ABA*) at the corners; three more isosceles (*DDE*) abutting the center; six scalene (*FCD*) completing the edges. Since this pattern is completed eight times to close the polyhedron, the 4ν octa has 128 triangular faces.

There is no reason to restrict this procedure to the octahedron. Diagram 5.9 shows a four-frequency icosahedron (4ν icosa). Since the icosahedron has twenty triangles to the octahedron's eight, each subdivided face has less curvature to supply and therefore can itself be more gently curved. Thus the different edge lengths are more nearly alike and the small triangles more nearly equilateral. The most distorted triangles, the corner ones, have side ratios of $0.447/0.320$ in the 4ν octa (1.4 to 1) and side ratios of 0.29 to 0.25 in the 4ν icosa (1.16 to 1). The greatest discrepancies in edge lengths (A and E) are $0.32:0.577$ for the octa (1.8 to 1) and $0.253:0.325$ for the icosa (1.28 to 1).

It is clear that an icosa yields less jerky polyhedra than an octa.

5.5

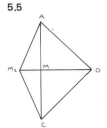

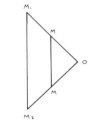

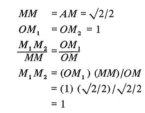

$$
\begin{aligned}
AO \quad &= CO = M_2O = 1 \\
AC \quad &= \sqrt{2} \\
AM \quad &= \sqrt{2}/2 \\
&= MO \\
M_2M \quad &= M_2O - MO = 1 - (\sqrt{2}/2) \\
AM_2 \quad &= \sqrt{AM^2 + M_2M^2} \\
&= \sqrt{(\sqrt{2}/2)^2 + [1 - (\sqrt{2}/2)]^2} \\
&= 0.765367
\end{aligned}
$$

$$
\begin{aligned}
MM \quad &= AM = \sqrt{2}/2 \\
OM_1 \quad &= OM_2 = 1 \\
\frac{M_1M_2}{MM} &= \frac{OM_1}{OM} \\
M_1M_2 \quad &= (OM_1)(MM)/OM \\
&= (1)(\sqrt{2}/2)/\sqrt{2}/2 \\
&= 1
\end{aligned}
$$

5.6

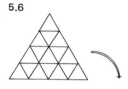

5.7

5.8

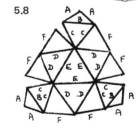

radius = 1

$A = 0.320$
$B = 0.447$
$C = 0.439$
$D = 0.518$
$E = 0.577$
$F = 0.459$

It is also clear that, octa or icosa, the higher the frequency, the smoother the arcs and the less the tiny triangles differ.

The acute reader will have been wondering what happened to our quest for Valence-4 polyhedra to generate Tensegrities. In the polyhedra we are learning to generate, six-way vertices preponderate, relieved only by twelve Valence-5 vertices when we start with the icosa, and by eight Valence-4 vertices when we start with the octa. Courage. The Tensegrity fan has a simple recourse, illustrated in Diagram 5.10. He has only to omit every other stripe in each of three directions on the face triangle, and lo, the Valence-4 vertices appear.

Following the usual Tensegrity rule of skipping every other vertex, we may insert 120 struts into six pairs of parallel lesser circles, ten struts per circle. The dip angle will be 180°/10, or 18°, and other dimensions may be computed as usual. We have only to remember that the radius yielded by Equation 3.5 is that of a *lesser* circle and divide it by the cosine of the lesser circle's "latitude" above the equator to obtain the radius of the actual sphere.

The surface pattern of the four-frequency Tensegrity icosa (T = 4ν icosa) is determined by twelve pentagons, each surrounding a vertex of the icosahedron from which we started (such vertices are always *surrounded*, not *occupied*, so we do not encounter the problem of devising a five-way Tensegrity intersection). Five triangles abut the five sides of each pentagon, and five hexagons touch its vertices. Since all twelve lesser circles are alike, all struts are the same length. The six different edge lengths into which we factored the icosa face in executing the original 4ν breakdown impose no complication because most of them disappear, and for Tensegrity purposes it is unnecessary to compute them at all. (This won't hold true of higher-frequency subdivisions.)

It is obvious that by extending this principle we can generate spherical Tensegrities as complexly multistrutted as we like. This means in practice that no matter how large a domical structure we may want to build, we can still use struts conveniently short and light. It will exhibit (assuming an icosa generator) a surface pattern of twelve pentagons per complete sphere, plus assorted triangles

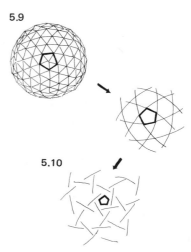

5.9

5.10

and hexagons. The triangles, increasingly numerous at higher frequencies, guarantee the stability of the structure. Lesser circles (and at certain frequencies great circles also) will run round the sphere in a three-way grid, and any circle we choose will supply a convenient plane for truncating the sphere to produce a dome.

We have only to remember that, for Tensegrity purposes, the frequency of icosa (or octa) subdivision must always be even, producing polygons of an even number of sides and permitting us to skip every other vertex when we insert the struts.

If it is not a Tensegrity that we contemplate, but some other kind of geodesic structure, the restriction to even frequencies does not apply.

6. Rigid Tensegrities

We are now ready to make the transition from Tensegrity structures to geodesic domes. We commence by noticing that the polyhedra we have been generating with the aid of frequency subdivision have one particularly useful property. As the frequency of subdivision increases and the number of edges grows more numerous, the surface of the spherelike form also grows smoother.

Analogously, in Tensegrity spheres, gaps and dips grow less and less obtrusive. We shall now see how this fact can be put to use.

Tensegrity polyhedra, we remember, are constantly interrupted by the sharp incisions of the dip angles. Whenever a running mouse comes to the end of a strut, it must leap the gap to the end of the next strut. How long a gap, we know from Equation 3.3. How deep a gap? How far down from its middle to the strut that passes below? This depth is a function of both gap and dip; as Diagram 6.1 shows,

$$\text{gap depth} = \sqrt{d^2 - (g/2)^2}. \qquad [\text{Eq. 6.1}]$$

But we know from Equation 3.3 and Equation 3.4 that gap and dip both diminish as the dip angle diminishes; and we know from Equation 3.2 that the dip angle diminishes as the number of struts around a great circle or a lesser circle increases. So increasing the number of struts around a circle — which is what we are doing when we increase the frequency — both shortens the gap and diminishes its depth.

This means (Diagram 6.2) that as we increase the frequency each gap narrows, and as it does so the strut that passes beneath it moves outward from the center, closer to the surface of the sphere. If it moves outward until the depth of the gap is just equal to the thickness of the struts, it will be so situated as just to touch

6.1

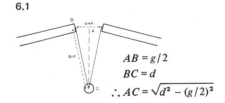

$AB = g/2$
$BC = d$
$\therefore AC = \sqrt{d^2 - (g/2)^2}$

6.2

the bottom edges of the upper struts (Diagram 6.3). By this time the gap will be extremely narrow, and a very little extension of the ends of the upper struts will ensure that they too will touch.

Let us suppose that the system is made of wood. We can cut the ends of the upper struts so they meet squarely, and we can bevel the upper edge of the lower strut so it fits snugly under them. We may also remember from Chapter 3 the option of replacing the large tensional diamond by a little sling that follows the lines of the dip. Diagram 6.3 shows it embedded in the wood, but it may easily be replaced by a couple of bolts. Tightening the nuts will draw everything together.

Now, note carefully: When we commenced this process, the rest of the system was tending to pull all three components apart. Now that they are touching each other, this is still true. The two upper struts are not in compressive contact: if released, they would pull *apart*. And they are not resting on the lower strut: if released, they would move outward and away from it.

The bolts and nuts are quite literally serving to *draw the elements together*. They are performing the function of the tension members in the original Tensegrity.

Repeat this throughout the domical structure, and we have an array of interlaced polygons, weaving alternately above and below the polygons they cross, and bolted at the crossing points. They outline pentagons, hexagons, and triangles and make a sturdy domical structure with no noticeable flexion (Diagram 6.4).

Having developed this structure from Tensegrity principles, we know about the tensile forces that are holding it together. Traceable through the bolts and the wooden beams, they form a tensional continuity, resembling the surface tension of a liquid sphere in being concentrated on the outer surface of the polyhedron.

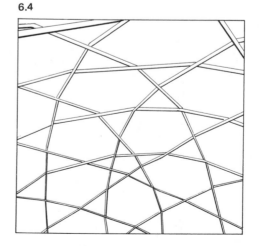

But a casual observer of the structure would soon commence studying the joints, and seeing what is diagramed in Diagram 6.3 he would suppose that the bottom member was bearing the weight of the upper ones. He would suppose that the structure was in *compression*, like a frame house, and that the bolts functioned merely as do carpenter's nails, to prevent lateral slippage. He would therefore form a wholly erroneous idea of the system's dynamics.

We have now reached the governing center of geodesic-dome design. The structure is lifted outward by a hidden tensional system. It resembles a contained explosion, like a balloon: compressive out-thrusts held in a tensile web, with this difference, that the forces pulling outward are also coming from the web. The structural members are not falling in together, nor in any important way leaning on one another.

However variant geometries and connecting systems may modify this account, it remains the first principle of geodesics.

Numerical Example

Some scratch-pad figures may help the reader visualize possibilities. An eight-frequency subdivision of the icosahedron (8ν icosa) is shown in Diagram 6.5, with the Tensegrity strutting indicated in heavy lines. Four struts span two triangles; since ten triangles girdle the icosahedron, the structural circles will be 20-gons, and the dip angle $180°/20$, or $6°$. Gap is 0.00274, dip is 0.026, and by Equation 6.1 gap depth is 0.026. (At high frequencies, gap depth is very nearly equal to dip). Radius is 4.79. Lesser circles A, A, being nearer the equator, will be slightly larger and will have longer struts than lesser circles B, B, but for the sake of a quick estimate we can assume an average strut length. Let's put it at 8 feet, a handy length in building with two-by-four lumber. We get a radius of 8 × 4.79, or 38 feet: say a 75-foot sphere. The gap will be a mere quarter of an inch. The interesting parameter is its depth, 8 × 12 × 0.026, or 2 1/2 inches. Since a so-called two-by-four is about 1 1/2 × 3 1/2 inches, two-by-fours turned edgeways will touch with an inch to spare for notching or beveling. We can bolt the whole thing together.

Buckminster Fuller erected such a framework about 1956, in Carbondale, Illinois. Since it followed a different geodesic geometry, it had fewer and longer members than those we have calculated, but the two-by-fours were bolted together, crossing under and over, just as we have described. It was very, very strong. Since

6.5

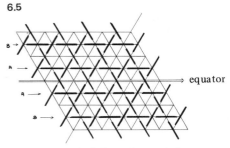

A, A, larger lesser circles
B, B, smaller lesser circles

Fuller had already by that time erected numerous geodesic domes of the familiar omnitriangulated variety, this 72-foot model constituted, as it were by afterthought, a demonstration that geodesics is in fact a special application of Tensegrity.

We now move on to geodesics proper: the generation of omnitriangulated, quasispherical, tensile-compressive free-span structures.

Part Two

geodesics

7. Great Circles

A geodesic dome is sliced from one of the complex polyhedra we were learning to design in Chapter 5. It has a large number of triangular faces, all approximately — but not quite — equilateral. The struts that bound them follow the paths of great circles, sometimes complete, more often interrupted. The cohesion of the whole, like that of a Tensegrity, is both compressive and tensile; it is helpful to imagine the tension system running along the outer surfaces of the struts, which are simultaneously loaded in compression.

As we saw in Chapter 1 (Diagram 1.1), tensile forces in post-and-beam construction run along the under edges of the beams, resisting central sag. In Tensegrities, on the contrary, tensile forces surround the system, hauling the vertices outward and so removing the need for posts. The first large geodesic structure (Robert W. Marks, *The Dymaxion World of Buckminster Fuller*, Carbondale: Southern Illinois University Press, 1960, figs. O.19-O.21) retained separate tension members, running to every vertex from pegs jutting from neighboring vertices; since then it has been clear that at moderate frequencies the tensile cohesion inherent in the upper surface of the strut has ample mechanical advantage, and in most domes the tensile members are inextricable from the compressive, concealing from the casual eye the crucial importance of tension.

It is evident that the hub systems, as well as the struts themselves, must resist pulling forces as well as pushing forces. When heavily loaded, either by its own weight or by an imposed stress, a geodesic system will tend to *pull apart* at some of the hubs.

Again, make a model before proceeding further. Diagram 7.1 shows a simple 3v. icosa; connections are flexible; clearly the strength is derived from geometry, not joinery.

7.1

five-way connector (use 12)
six-way connector (use 80)

Repeat face triangle 20 times to make the sphere

With 3/8-inch connector:
A = 9.7 cm. (use 60).
B = 11.3 cm. (use 90).
C = 11.6 cm. (use 120).

With pins or paper fasteners:
A = 10.7 cm.
B = 12.3 cm.
C = 12.6 cm.

Experiment.

Try removing one strut.

Try removing one connector.

Note that at exactly twelve points, evenly distributed about the sphere, five triangles meet at a vertex. Elsewhere, the number that meet is always six.

Make an icosahedron (Diagram 7.2) with twelve more connectors and twenty struts of equal length. Note the five triangles meeting at each vertex. Observe the resemblance to the 3v icosa sphere.

Finally, observe the circle patterns. The five-way vertices are called pents *(after the pentagons that surround them). From each pent center radiate portions of five great circles. Each has its center at the center of the system. Each sets off on a 360° circuit, of which it completes about 63.5° before bumping into another pent center (Diagram 7.3).*

But if you follow the lead of any pentagon edge, you can trace a circuit clear around the sphere. This is a lesser circle; its center is not at the center of the system. They girdle the sphere in parallel pairs (Diagram 7.4), like the Tropic of Cancer and the Tropic of Capricorn.

It is natural to think of cutting the sphere along one of these lesser circles to get a tall or a shallow dome. Try it by unfastening connectors. Your first discovery will be that the dome does not sit quite flat: the struts in the lesser circles are almost, but not quite, coplanar. Obviously, the lengths of struts leading to the lesser circle could be adjusted to get a level truncation plane. *That's something design must take into account.*

The next discovery to be made from a dome model is that the perimeter wobbles. Away from the perimeter, the structure resumes its rigidity. As in a Tensegrity, tensile circuits must be completed. A dome design is not completed until its lower edge is fastened to the earth. It is ultimately the inelasticity of the earth itself that the tensile network transmits through the structure.

The next step in investigating domes and great circles is to make a companion model based on the octahedron instead of the icosahedron. Diagram 7.5 gives data. Comparing the two models, we notice several things immediately.

7.2

7.3

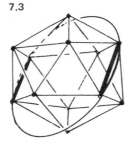

interrupted great circle
(2 opposite edges)

7.4

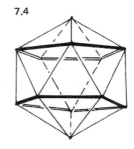

parallel
lesser circles

● *The octa uses fewer struts and connectors, and has fewer faces.*

	3ν Icosa	3ν Octa
edges	270	108
vertices	92	38
faces	180	72

● *The octa is much less smoothly spherical. This derives from the fact that an octahedron is a poorer approximation to a sphere to begin with. Its sphericity can be improved by raising the frequency. A 5ν octa has a component inventory comparable to that of the 3ν icosa (300 edges, 102 vertices, 200 faces) and is comparably smooth in curvature.*

● *Instead of the icosa's twelve pents, we find six squares, coresponding to the six vertices of the parent octahedron.*

● *Like the icosa's great-circle arcs from pent center to pent center, the octa has great-circle arcs running from square center to square center. But instead of being intercepted at the next square center, each meets a partner at 180° and in effect continues straight through. Thus every octa, regardless of frequency, has three mutually perpendicular great circles outlined by its struts, making natural hemispherical truncation planes. (Icosas display hemispherical great circles at all even frequencies.*)*

● *Like the 3ν icosa, the 3ν octa also displays wobbly lesser circles. Truncation at other than hemispheric intervals will require adjusting strut lengths to make one of these lesser circles planar.*

It is clear that certain characteristics of the parent polyhedron carry over into its geodesic derivates. The icosahedron has no vertex-to-vertex great circles, and neither has any of its geodesic subdivisions. The octahedron on the other hand has three, and so have its breakdowns. The icosahedron has twelve vertices where five triangles meet, the octahedron six vertices where four triangles meet; the same is true of their frequency subdivisions. (In either case, all other vertex points are hexagonal, by-products of the three-way grid.) Most important, an observer at the center of the

*These do not run vertex-to-vertex but are by-products of the subdivision system, which at even frequencies cuts triangles in half.

GREAT CIRCLES   49

7.5

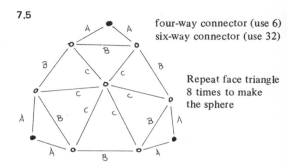

four-way connector (use 6)
six-way connector (use 32)

Repeat face triangle 8 times to make the sphere

With 3/8-inch connector:
A = 13 cm.   (use 24)
B = 18.3 cm. (use 36)
C = 19.5 cm. (use 48)

With pins or paper fasteners:
A = 14 cm.
B = 19.3 cm.
C = 20.5 cm.

system would look in the same directions to see the same pent points (icosa) or square points (octa), regardless of frequency. We shall later be deriving important consequences from this fact.

Great-Circle Systems

We can now look more closely at the relationship between polyhedra and great circles. The most vivid look is prepared for by making an icosahedron of cardboard (twenty equilateral triangles, about 2 1/2 inches to a side, glued together at the edges with contact cement).

Press fingertips to a pair of opposite vertices and turn the icosahedron on the axis your fingers define. If you now imagine it spinning rapidly, you can see where the equator of spin would lie. If you next move the fingertips to a new pair of vertices without changing the icosahedron's orientation in space, it acquires a new axis and its spin will trace a new equator, inclined to the first. If you keep this up till you have used all six pairs of opposite vertices, you will have six equators symmetrically intersecting one another. Diagram 7.6 shows all of them and shows how they subdivide symmetrically the faces of the spherical icosahedron. This is the family of the *six great circles*.

We next repeat this exercise, using midedge points for the poles of spin. There are fifteen pairs of opposite edges, and rotating the icosahedron between fingers pressed to the midpoint of any pair will describe an equator like the one in Diagram 7.3, cutting through two pairs of triangles and coinciding with two edges. By the time we have used all the axes of rotation we have located the *fifteen great circles*. Diagram 7.7 shows how they outline every icosahedral face, and also slice through every face three ways.

There is one more great-circle family, which we locate by using midface points for the poles of spin. (A triangle's midface point is located by joining each vertex to the midpoint of the opposite side. It is two-thirds of the way down each altitude and is also called the triangle's center of gravity.) The icosahedron has 10 pairs of opposite faces, and spinning it successively on an axis passing through each pair gives us the *ten great circles*. Diagram 7.8 shows how they are located. If we follow any one of them along

7.6

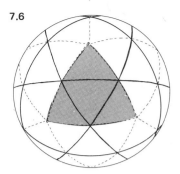

SIX GREAT CIRCLES
(Broken lines = icosa edges.)

7.7

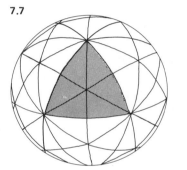

FIFTEEN GREAT CIRCLES

7.8

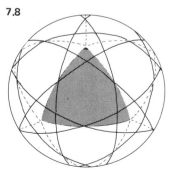

TEN GREAT CIRCLES
(Broken lines = icosa edges.)

its successive crossings of icosa triangle edges, we shall see that it repeatedly leaves the midpoint of an edge to cross the next edge at right angles.

So the icosahedron has three families of symmetrically placed great circles, totaling 6 + 15 + 10, or 31.

We can now try some combinations. Start with the six great circles. Next outline the faces of the spherical icosa; we do this by using *portions only* of the fifteen great circles. Diagram 7.9 shows the result: a *2v* geodesic breakdown. If the vertices were joined by chords instead of arcs they would diagram a perfectly useful structure. They would be of two different lengths, and the longer ones would follow the six great circles. The shorter members, outlining icosa edges, define portions of interrupted great circles. The subdivisions tend to run in the same directions as the icosa edges, and we see at once why any *even*-frequency subdivision of this general design will yield equators: they are the six great circles.

We now try something else. Commence with the fifteen great circles, and this time remove the portions that outline icosa edges, leaving everything else. This gives us (Diagram 7.10) another *2v* breakdown, this time with no *complete* great circles at all. (By adding four members, to complete two icosa edges, you could make an equator that coincided with any one of the fifteen great circles. The new members would cut four triangles in half.) This kind of breakdown, cutting perpendicularly across icosa edges that are no longer present, is called Class II. (The breakdown discussed first is Class I.)

We may remember that the ten great circles also cut icosa edges perpendicularly, and try inserting them. Diagram 7.11 shows why this is unsatisfactory. It approximates a *4v* breakdown, but the triangles are of markedly different sizes, and some of them are excessively isosceles.

Still, we can understand why, in the pioneer days of geodesics, this sort of subdivision was the first to be explored: it promises access to the largest inventory of great circles, potentially twenty-five. Breakdowns of this general configuration were at first called regular, later triacon (after the rhombic triacontahedron, whose contours it suggests). It proceeds by joining each vertex of a tri-

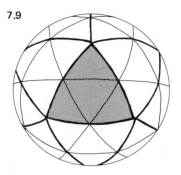

7.9

2*v* BREAKDOWN
(Six great circles + portions only of fifteen)

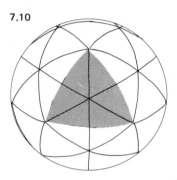

7.10

2*v* BREAKDOWN
(Portions of fifteen great circles)

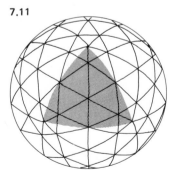

7.11

4*v* BREAKDOWN
(10 + 15 = 25 great circles, none complete)

angular icosa face to the midpoint of the opposite edge (which gives portions of fifteen great circles) and running additional lines of subdivision more or less parallel to these (which with luck will pick up portions at least of ten more). In the finished artifact only this grid remains: the icosa edges themselves disappear. Later investigation produced the method that yields subdivisions parallel to triangle edges. Being an alternative to its predecessor it was called the alternate breakdown, a name that has stuck, though "Class I" is more formal.

In what we may call the prehistory of the art, there were no breakdown systems at all. Buckminster Fuller's first experimental domes were simply sets of icosahedral great circles, made of short chordal members bolted together. All thirty-one great circles made a structure of enormous strength, and they intersected one another so frequently that the maximum component length was never excessive (about 9 1/2 feet for a 50-foot-diameter structure). Still, disadvantages of the structure strike the eye: component lengths vary over a range of more than two to one, and so does vertex population. Some vertices receive only four members; others are cluttered with ten and twelve.* It is intuitively obvious that such a system contains much redundancy and (putting the same thing differently) a good deal of nonfunctional weight.

Omitting six great circles eased the clutter somewhat; there were families of structures with twenty-five great circles. The next step, obvious in retrospect, was perhaps the most difficult to conceive: *interrupting* fifteen of the twenty-five great circles by removing members that outlined icosa faces (Diagram 7.10), and extending members that had terminated on an icosa edge to reach into the next triangle. All vertices were then sixes and fives, and strut-length variation was just under 2 to 1.

The system was now about to take its leave of what seems to us, decades later, a naive dependence on inherent great circles. (The name *geodesic dome* preserves the memory of that initial dependence.) It is obvious that more members can be introduced, parallel to the ones we have. This gives rise to the concept of frequency subdivision. It also permits the shorter component lengths we shall need for very large structures. It is also obvious

*See Marks, Fig. O-4.

that by tinkering with the geometry we should be able to bring strut lengths closer to equivalence. The systems of geodesic breakdown commence at this point.

Octahedral and Tetrahedral Great Circles

The octahedron too has three families of great circles, which the same experimental method will locate. The three vertex-to-vertex axes yield *three great circles,* which coincide precisely with the octahedron's edges. Midedge to midedge axes yield *six great circles,* of which three appear on any single face, coinciding with its medians. The axes passing through opposite pairs of midfaces yield *four great circles,* of which any face exhibits parts of three, joining the face's midpoints. So the octahedron has 3 + 6 + 4 = 13 great circles, of which any face is bounded and crisscrossed by 9.

If we design a geodesic sphere or dome by subdividing an octahedron, this set of thirteen comprises all the *complete* great circles we can possibly obtain. All others will be interrupted. Any Class I subdivision yields three, the ones bounding the octa edges. (Hence all Class I octa subdivisions have natural equators.) Any *even-frequency* Class I octa subdivision has these three and also four more (look back at Diagram 5.2). So all even-frequency Class I octa subdivisions contain seven complete great circles each. But as long as the lines of subdivision run parallel to the octa edges, there is no way to incorporate the remaining six great circles at all. Class II subdivision gives us access to these, but loses the other seven.

The tetrahedron, in this as in many other respects, is bafflingly economical. It has no opposite pairs of vertices. Each vertex is opposite a midface, and joining these unmatched pairs by axes we obtain four equators of spin, *four great circles,* of which any single face exhibits parts of three. Axes through pairs of midedge points yield another *four great circles,* of which each edge again exhibits parts of three, joining its midedge points. That is all.

An even-frequency Class I subdivision will display the first set, and if the frequency is divisible by 4 the second set will be present, as well. Class II subdivisions with frequencies divisible by 4 display the second set of great circles if they follow Method 1 (see Chapter 10) but none at all if they follow Method 3.

8. Symmetry Systems

Symmetry is a measure of a system's ability to absorb rearrangements that cannot be detected. A jigsaw puzzle has none; if you try to interchange two pieces, fit is destroyed. By contrast, square tiles in a square box can be permuted indefinitely without anyone's noticing. An equilateral triangle which we are forbidden to turn over can be placed on the table three ways — any vertex upward — with no likelihood of these states' being distinguished.

When we turn a polyhedron on an axis, as we did when we were eliciting the great-circle families in Chapter 7, it passes through a sequence of successive states that cannot be told apart. If the icosa axis runs vertex-to-vertex, a complete rotation brings five successive aspects to the fore, each indistinguishable from its predecessor. Each midedge axis yields two indistinguishable positions, and each midface axis three. So the three great-circle families correspond to three groups of rotational symmetry, and it is not surprising that we can discover symmetries in the interrelationship of the great circles themselves. In fact, the icosahedron's thirty-one great circles can be completely specified with the help of a small triangular module, shown in Diagram 8.1. Systematically rotated and reflected six times per spherical icosa face (thus, 6 × 20 = 120 times per sphere) it reconstitutes the entire great-circle system. This little triangle thus *contains all the information we need to subdivide a sphere with symmetrical great-circle systems.* Any further subdivisions we may execute within the symmetry triangle will likewise be propagated throughout the system. The symmetry triangle is thus a kind of geodesic DNA molecule, and so long as we are concerned with spheres we may confine our interventions to its minute terrain, in confidence that the rest of the system will reproduce them exactly.

The octahedron has a corresponding symmetry triangle, which

8.1

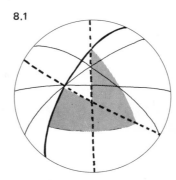

GREAT-CIRCLE FAMILIES

——————— ₁₀ SYMMETRY TRIANGLE (blank)
——————— ₁₅ CONTAINS A SAMPLE OF ALL
- - - - - ₆ GREAT-CIRCLE RELATIONSHIPS.

covers not 1/120 but 1/48 of its spherical area. Inspecting a spherical tetrahedron likewise, we discover a symmetry triangle that comprises 1/24 of it. Since it can be shown that these three polyhedra — icosahedron, octahedron, tetrahedron — contain the entire available array of great-circle symmetry systems, there are no other cases we need consider.

It is helpful to think of the polyhedron simply as *a device to help us find systems of symmetry.*

You are standing at the center of a polyhedral structure. You observe the repeating faces, the uniformly radiating edges, the equidistant vertices. These locate the reference points of omnisymmetry. A vertex, a face center, a midedge, comprises their minimum set. The little right-angled triangle so marked out — the symmetry triangle — defines all the polyhedron's capacity for changes-that-do-not-make-any-difference. Its corners are the poles of symmetry, and if we could locate and specify them somewhere on a single polyhedral face, we should have control of the whole omnidirectional array. To see how this might be done, we shall imagine sightlines from the center of the system and find an economical way to specify them.

The great sphere of the sky, a man-made geodesic sphere — it doesn't matter which we think of: the sightlines are equivalent. The sky has one advantage: it frees us from error in locating the center. You are always at the center of the celestial sphere.

We can't see the whole sphere, but it suffices to see the upper half of it. Imagine an unimpeded horizon, up from which arches the hemispherical bowl. One point on this bowl is perfectly unambiguous, the point directly overhead and equidistant from every point on the horizon. Call it 0,0.

We can now designate any direction we choose with the aid of just two numbers. The first number is a compass direction: zero if we are facing north, an increasing number of degrees as we turn counterclockwise from north. The second number also specifies degrees: zero at the zenith, 90° at the horizon, other figures in between. Thus 0,45 designates a unique direction: north, and halfway from zenith to horizon. The corners of the octahedral symmetry triangle are at 0,0; 0,45; 45,54.7356[+] (see Diagram 8.2;

8.2

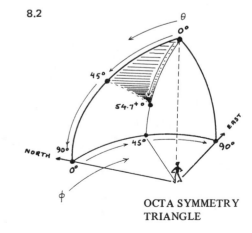

OCTA SYMMETRY
TRIANGLE

the derivation of this last figure will be explained later).

Within the symmetry triangle we may specify in the same way additional points, each with its pair of numbers. We can choose them so that they locate whatever great circles we intend to use, and the vertex points for the geodesic breakdown we plan to employ. Each point designates a sightline, and what we shall have specified is a cluster of sightlines which *contains all the information we need to build a geodesic structure.* The symmetry triangle and the cluster of points it contains would simply be reproduced, over and over, until the whole hemisphere was contained — or the whole sphere, for that matter.

If you were standing at the center of a man-made geodesic hemisphere you would see the vertex points these sightlines designate, all joined by structural members in triangular arrays. With a little study you could locate the boundaries of an enclosing symmetry triangle and verify that it was the module of repetition.

Envelope Contours

We are now about to derive advantage from the fact that we have directed our reference sightlines at the sky instead of at something more accessible like a planetarium ceiling. The advantage inheres in the fact that the sky has no definite distance. Celestial locations are nothing but a set of distanceless directions. Despite the three-dimensional ambience, two coordinates, not three, will suffice to locate any star, any point, any polyhedral vertex. Big or small, any icosahedral or any octahedral system disposes its reference points along the same sightlines from the center. We can therefore treat the whole system as a set of directions and invoke the distances later when we need them.

It's also clear that the *contour* of the bounding surfaces makes no difference to polyhedral sightlines. The fact that some stars are much further away than others does not affect their relative placement in the sky. In the same way, *a geodesic envelope may be of any shape without affecting the direction in which we sight to look at one of its vertices.*

Provided the structural members were thin enough for their

apparent width to provide no cues, it would be quite impossible
for an observer at the center of a man-made structure to tell
whether he was surrounded by a geodesic sphere or a geodesic egg.
In the former case his eye's distance from every vertex would be
the same, in the latter case it would vary, but, given a structure so
refined its members looked like ruled lines, these two cases would
display exactly the same angular symmetries.

In separating *distance* from *direction*, we have taken a step of
great potential usefulness. First, we know better what we shall be
about when we execute a geodesic breakdown. The breakdown
will be a little cluster of sightlines, a local symmetry occupying
one symmetry zone of the great polyhedral symmetry, and repeated as often as the structure we envisage may require. The envelope
of that structure will intercept our system of directions at various
points. These are the points that will be joined by struts, and the
length of the struts will depend wholly on the size and contour of
the dome we want to make. Specify these, and from a table of
angular sightlines it is simple to calculate the distances.

It is evident from Diagram 8.3 that if α is the angle between the
sightlines that lead to two vertices, and r is the distance of each
vertex from the center (the uniform radius, in the case of a spherical system), then the strut length between the points is r multiplied by $2 \sin (\alpha/2)$, twice the sine of half the angle. Twice the
sine of half the angle is called a *chord factor*, because the strut is a
chord and we have here one of two useful factors of that chord,
the other being the radius. Geodesic designers have worked with
chord factors ever since Fuller introduced the concept in the early
1950s. Tables of chord factors, containing as they do the essential
design information for spherical systems, were for many years
guarded like military secrets. As late as 1966, some 3ν icosa
figures from *Popular Science Monthly* were all anyone outside the
circle of Fuller licensees had to go on. Now, thanks largely to
Lloyd Kahn's *Domebook 1* (Los Gatos, Ca., Pacific Domes: 1970)
and *Domebook 2* (Bolinas, Ca., Pacific Domes: 1971) chord
factors for at least small-to-medium structures are fairly accessible.

But concentration on chord factors — the information a con-
structor ultimately wants — has concealed the great usefulness of

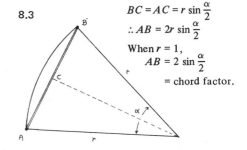

8.3

$$BC = AC = r \sin \frac{\alpha}{2}$$
$$\therefore AB = 2r \sin \frac{\alpha}{2}$$
When $r = 1$,
$$AB = 2 \sin \frac{\alpha}{2}$$
= chord factor.

the angles from which they are derived. If we know the chord factors for a sphere of a given breakdown and frequency, we can proceed to duplicate it in any practical size. But if we know the sighting angles, we are no longer restricted to spheres. By allowing the radius to vary harmoniously from point to point, we can derive chord factors for a dome of any useful shape. There is no need whatever for geodesic structures to be spherical.

What we are after, therefore, is a local cluster of symmetrically disposed points, their symmetries expressed as angles between sightlines from a common center. Once we have our cluster of points, which will be denser for higher-frequency breakdowns, and have made a table with sighting angles for each, we shall need only a procedure for determining the chordal distances from any point to its neighbors, whether they are at the same distance outward or not, and we shall be equipped to design geodesic structures of any shape we please.

9. The Spherical-Coordinate System

We have learned that two angles — call them ϕ and θ — will specify a direction, and two angles ϕ, θ, together with a distance r will specify a unique point. These are the elements of the *spherical-coordinate system,* which is less well known than it might be. Most problems of geodesic mathematics can be solved with its aid with straightforward ease.

The ϕ coordinate resembles a meridian of longitude. All points with the same ϕ lie on the same great circle, which passes through the zenith (0,0). It is convenient to incorporate the reference meridian, $\phi = 0$, into our calculations whenever we can. The first coordinate of all points situated on it will be 0. The θ coordinate resembles a specification of latitude, except that it is measured down the ϕ circle from $0°$ at the zenith, instead of upward from the equator. (In the spherical-coordinate system, the equator is at $\theta = 90°$.) All points whose second coordinate θ is 90 lie on the equatorial great circle. With the exception of these, all points with the same θ lie on the same *lesser* circle and specify a plane along which the dome can be truncated so it will sit flat.

The r coordinate, measured not in degrees but in units of length from the center, tells us how far out a point is to be found. If the points are on the surface of a sphere, like the vertices of a spherical geodesic structure, we can let $r = 1$ and virtually disregard it. The angles ϕ and θ tell us all we need to know. If the envelope of the system is other than spherical, then varying values of r will have to be taken into account.

Now for the usefulness of understanding this system. If we know the coordinates ϕ_1, θ_1, r_1 and ϕ_2, θ_2, r_2 of two points, then we can find the distance between them by inserting these values into an equation which a pocket calculator can readily subdue.

The result, d, will be in the unit used for r_1 and r_2, and if either of these distances — or both in the case of a sphere — is made equal to 1, the result will be a chord factor.

$$d = \sqrt{r_1{}^2 + r_2{}^2 - 2r_1r_2 \left\{\cos \theta_1 \cos \theta_2 + \cos (\phi_1 - \phi_2) \sin \theta_1 \sin \theta_2\right\}}.$$

[Eq. 9.1]

For the spherical case, when $r_1 = r_2 = 1$, this becomes

$$d = \sqrt{2 - 2 \left\{\cos \theta_1 \cos \theta_2 + \cos (\phi_1 - \phi_2) \sin \theta_1 \sin \theta_2\right\}}.$$

[Eq. 9.2]

For a first example, check back to the 2ν octahedron we discussed in Chapter 5 (Diagram 5.4). Diagram 9.1, with the same lettering, shows how this looks in spherical coordinates. A, B, C, are the corners of one octahedral face. Their ϕ, θ coordinates are 0,0; 0,90; 90,90. The edge midpoints M_1, M_2, M_3 have the coordinates 0,45; 90,45; 45,90. Because the system is symmetrical we need only two distances: a vertex to a midpoint, and a midpoint to a midpoint. A-M_1 and M_1-M_2 will suffice. Because the system is spherical we use Equation 9.2. The sines and cosines we shall need are as follows:

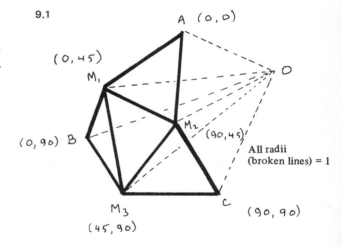

9.1

	sine	cosine
0°	0	1
45°	0.70710678	0.70710678
90°	1	0

Inserting these values into Equation 9.2, we get for A-M_1 (coordinates 0,0 and 0,45):

$$d = \sqrt{2 - 2\,(0.70710678 + 0)}$$

$$= \sqrt{2 - 1.41421356}$$

$$= \sqrt{0.58578644}$$

$$= 0.765367$$

And for M_1-M_2 (coordinates 0,45 and 90,45) we get

$$d = \sqrt{2 - 2(0.5 + 0)}$$

$$= \sqrt{1}$$

$$= 1.0$$

These answers agree exactly with those we obtained by other means in Chapters 5 (Diagram 5.5). And since they are multiples of r when $r = 1$, they are identical with the chord factors introduced in Chapter 8.

True, the purely geometrical argument in Chapter 5 yielded the chord factors with less manipulation and in less time. But that was a $2v$ case. Raise the frequency, and the geometrical method soon becomes too cumbersome to stay in the running. (Desert the spherical contour, and it becomes nearly impossible.) Moreover, each result yields data for its successor, and an early error invalidates subsequent work. The ϕ, θ method, on the other hand, permits a standardized procedure. All chord factors are found by exactly the same routine, and all are found independently; errors are isolated, and do not affect other computations. And a good pocket calculator, which can quickly change angles into their sines and cosines and has no more trouble with roots than with $2 + 2$, permits us to run through the chord-factor equation for each pair of points in less than a minute, even when coordinates are cumbersome. (Suitably programmed, the ultrasophisticated HP-65 will do the computing in four seconds per case.)

While we have in front of us data for a simple case, we may as well experiment with Equation 9.1. What if we wanted to make this simple dome elliptical? Let us specify a beehive orientation (Diagram 9.2) 1 1/2 times as high as it is wide. A hemisphere of 20-foot diameter would thus be stretched to give not 10 feet of headroom but 15.

We see at once that the radius varies with θ. At the zenith ($\theta = 0$), $r = 1.5$. At the floorline ($\theta = 90°$), $r = 1.0$. At the octa midedge ($\theta = 45°$), the reader may take it on trust that $r = 1.1767$. (We shall learn later how to calculate r for any value of θ in ellipses of any proportions.) These are the only values of θ we have to deal with, so we make a $\theta : r$ table:

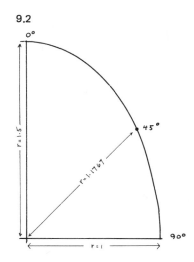

9.2

$$
\begin{array}{cc}
\theta & r \\
0° & 1.5 \\
45° & 1.1767 \\
90° & 1.0
\end{array}
$$

Since r continues to vary beyond the confines of the octa symmetry triangle, we should identify the various vertices in a face triangle (Diagram 9.3 shows a useful notation) and decide how many different chord lengths we must deal with. 20/21 and 21/22, where $r = 1$, will be just the same as before. But since 0,0 and 1,0 have special values for r, we shall need 10/11, 00/10, 10/20, and 10/21. But the system retains mirror symmetry (1,0 has the same θ as 1,1), so 00/11 will equal 00/10, 10/20 will equal 11/22, and 11/21 will equal 10/21. So we need four new chord factors, which we obtain by putting the proper values of ϕ, θ, and r into Equation 9.1. Thus 10/21 is given by:

9.3

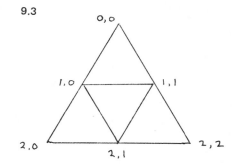

$$
d = \sqrt{1.1767^2 + 1^2 - 2 \times 1.1767 \times 1 \,(0.70710678 \times 0.70710678)}
$$

$$
= \sqrt{1.2079}
$$

$$
= 1.099.
$$

The reader may verify that 00/10 is 1.067, 10/20 is 0.8488, and 10/11 is 1.1767.

A model built to these specifications would be far from smooth — the frequency is simply too low to record nuances of elliptical curvature — but the few points we have are correctly located, and it is clear that the principle of elliptical domes presents no real difficulties.

So we know what to do with coordinates of breakdown vertices on a polyhedron face. Next we need to know which polyhedron to choose, and how to break it down to obtain the coordinates themselves.

10. Breakdown Systems

We start with the breakdown systems, because they provide some of the criteria for choosing a polyhedron. All systems start with a triangular polyhedron face and subdivide it with a three-way grid. We then push all the vertices of this grid outward till they are a common distance from the center (for spheres) or a harmoniously varying distance (for elliptical and other contours). When we have done this, some of the lines of subdivision will follow great circles inherent in the polyhedron's symmetry. We want the ones that don't do this to trace at least partial great circles, interrupted by the edges of polyhedron faces. And we want to maintain a reasonably symmetrical spacing, so that the little triangles we have created will be nearly equilateral. We shall commence our thinking about all this by examining the first of the two basic ways to orient the grid.

Class I (Alternate)

The grid lines run roughly parallel to the edges of the polyhedron face.

We want to specify the crossing points, and there are a number of different techniques, all of which yield slightly different strut lengths. All are correct, in the sense that they generate three-way grids with sphere vertices equidistant from the system center. Some are of historical interest chiefly. Still, there are choices to be made, and particularly at low frequencies we may be guided between two principal methods by a rather simple choice: similarity of triangle *size,* or similarity of triangle *shape.*

A worked example will clarify this. To keep figures few and simple we shall use the octahedron and compare two methods of making a 3ν Class I breakdown.

Diagrams 10.1 and 10.2 show Method 1. It amounts to drawing a triangle face on flat paper, dividing its edges equally, then projecting the division points outward to obtain equal radii. It is clear from Diagram 10.2 that although the divisions *AC, CD, DB* are equal, the chords *AF, FG, GB* are not (the end chords are shorter than the central one). The sighting angles from the center to the chord ends will differ likewise, so the θ coordinates will be unevenly spaced. Diagram 10.2 shows a construction for obtaining them. As for the triangle's center point, we have learned that it is 54.7356° inward from each vertex (coordinates 45, 54.7356). Equation 9.2 will give us chord factors as shown.

Method 2* simply spaces the θ coordinates equally down the octa edge (30°, 60°, 90°), making all edge struts equal. The triangle's center is located as before, and we proceed (Diagram 10.3) to obtain chord factors.

We should now look at models, and folding paper is an easy way to make them. Diagram 10.4 and Diagram 10.5 show patterns for one octa face computed by each method. Four of either, cut out and creased, can be taped together to make a hemispherical dome. By simply looking at the diagrams we can see a marked difference. When we have equal chord factors along the octa edge (Method 2), the small triangles vary less in size, but vary more in shape. Disregarding for the moment the corner triangles *A*, which are always a special case, we notice that the rest of the octa face is shared by three each of two quite noticeably isosceles triangles, *B* and *C*. But when we vary the edge chord factors (Method 1) the corner triangles *D* stay virtually the same shape as before, while the rest of the octa face divides itself into 6 identical triangles *E*, which come much closer to being equilateral than either *B* or *C* in the other diagram.

In short, *by letting the triangle sizes vary more we can keep their shapes more nearly constant.*[†] This seems structurally as well

---

*The class and method terminology is that of Joseph Clinton, "Geodesic Math," *Domebook 2*, pp. 106-107.

[†] A note in *Domebook 2* (p. 112) set out to make this point but had the misfortune to get it exactly reversed. The "smoother arcs" and the less equilateral triangles go with Method 2, not Method 1, as investigation of the numbers in the *Domebook* tables will show.

10.1

10.2

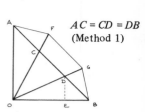

AC = CD = DB
(Method 1)

$AB = 1$ $DB = CD = AC = 1/3$ $OB = 1/\sqrt{2}$
$DE = EB = 1/3 \sin 45°$
$OE = OB - EB = 1/\sqrt{2} - 1/3 \sin 45°$
$\therefore \angle DOB = \arctan (1/3 \sin 45°)/(1/\sqrt{2} - 1/3 \sin 45°)$
$\qquad\qquad\qquad\qquad = 26.5650512°$
$\angle COD = 90° - 2\angle DOB = 36.8698977°$
O   = Octa Center
AB   = Octa Edge
AFGB = Projection onto sphere

method 1 coordinates		chord factors	
	ϕ	θ	
0,0	0	0	0,0/1,0	0.459506
1,0	0	26.5650512	1,0/2,0	0.632456
1,1	90	26.5650512	1,0/1,1	0.632456
2,0	0	63.4349488	2,0/2,1	0.671421
2,1	45	54.7356103		

10.3

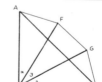

$\angle x = \angle y = \angle z = 30°$
$\therefore AF = FG = GB$
(method 2)

method 2 coordinates		chord factors	
	ϕ	θ	
0,0	0	0	0,0/1,0	0.517638
1,0	0	30	1,0/1,1	0.707107
1,1	90	30	2,0/2,1	0.650115
2,0	0	60		
2,1	45	54.7356103		

as aesthetically preferable. The more nearly equilateral a triangle, the more nearly equilibrated should be its resistance to pushes and pulls from various directions. The price, a more marked variation in strut length, has been carried to extremes by the present example. It is less marked at high frequencies than at low, and less for the icosa than for the octa. A $3v$ octahedron is about the worst possible case.

Since Method 1 appears to be the Class I breakdown of choice, we shall stick to it for the rest of the book.

A word about triangle shapes in general. Though the norm is an equilateral triangle, it is clear that if the little triangles were really equilateral we should have neither dome nor sphere but flat gridwork. It is their departure from equilaterality that provides the curvature, as we can see by considering the sum of the angles around a single vertex. Around the central vertex in Diagram 10.5, the apices of the E triangles total $6 \times 56.2°$, or $337.2°$. This is nearly $23°$ short of the $360°$ that would lie flat; the shortage is called the *spherical excess* of the vertex. Descartes showed long ago that the spherical excess of all the vertices in any polyhedron must always total $720°$. Since there are 38 vertices in a $3v$ octa sphere, we should find an average spherical excess of nearly $19°$ at each of them. The vertex we have just looked at has more spherical excess than the average, and so have seven others like it; they are contributing to a deficit incurred at the vertices where four octa faces meet, and where $4 \times 86.9° = 347.6°$ gives a spherical excess of barely more than $12°$.

It is at these latter vertices — the octa's six squares, the icosa's twelve pentagons — that we are most likely to find deficient spherical excess, in other words a tendency to flatness. Flattish vertices are more likely to pop in under load, one reason it is desirable to keep spherical excess more or less evenly distributed (and triangles more or less equilateral).

Class II (Triacon)

The grid lines run roughly perpendicular to the edges of the polyhedron face.

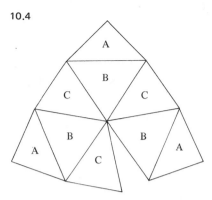

10.4

3v OCTA
METHOD 2, EQUAL-EDGE DIVISIONS

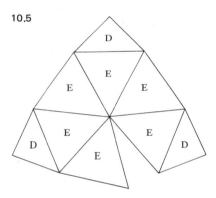

10.5

3v OCTA
METHOD 1,
EQUAL CHORDS BEFORE PROJECTION

This breakdown has a reputation for being difficult to understand. To set about demystifying it we may note that it does not really break down the icosa or octa face from which it derives its frequency count, but a smaller and nonequilateral triangle which consists of two icosa or octa symmetry triangles back to back.

It arose as a method of incorporating the great circles that coincide with face-triangle medians. If we draw the three medians of a face we outline the symmetry triangles (Diagram 10.6). If we draw the medians of two adjacent icosa faces (Diagram 10.7) we find that we have outlined a pair of triangles, *ADE* and *DEC*, each comprising two symmetry triangles. If we continue in this way to subdivide the entire icosa (Diagram 10.8) and then eliminate the icosa edges themselves, we have a 2*v* Class II breakdown, made up entirely of triangles like *ADE*.

If we now attempt further subdivision of the icosa face by lines parallel to the medians we first drew, we discover that only even-frequency subdivision is possible. Diagram 10.9 shows 4*v*, and Diagram 10.10 shows that what we have really achieved is a 2*v* subdivision of the triangle *ADE*. Since in Diagram 10.8 we had covered the entire sphere with triangles like *ADE,* such a triangle is the real module of Class II subdivision. There are sixty such triangles, clustered five to an icosa vertex, and they are what the Class II subdivision really subdivides. Since they seem to have no geometrical name, we may as well call them Class II triangles. (The diamond *ADEC,* made up of two of them, is one face of a spherical rhombic triacontahedron, hence the nickname *triacon* for this breakdown.)

Once more there are several methods, of which we shall find uses for two.

Method 1 in effect divides the diamond median (= icosa or octa edge) exactly as in Class I, Method 1, to yield equal parts on flat paper but unequal chords when the vertices are pushed out to make spherical radii equal. We then concentrate on one half of the diamond, the Class II triangle, and run perpendiculars from the median outward to the sides. The triangle's bottom edge is now subdivided the way its median was, and lines join these edgepoints to complete the grid.

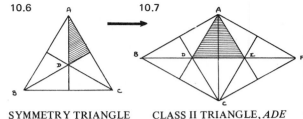

10.6   10.7

SYMMETRY TRIANGLE   CLASS II TRIANGLE, *ADE*

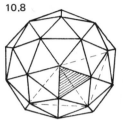

10.8

2*v* CLASS II SPHERE
(2 ICOSA FACES, IN BROKEN-LINE OUTLINE; CLASS II TRIANGLE, SHADED)

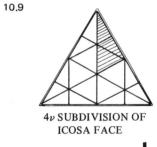

10.9

4*v* SUBDIVISION OF ICOSA FACE

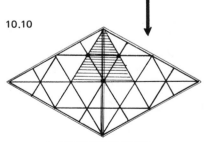

10.10

2*v* SUBDIVISION OF CLASS II TRIANGLE

The result is a strikingly symmetrical breakdown, and a little more economical of components than Class I, but not much. Its symmetry module is the icosa or octa symmetry triangle — that is, half a Class II triangle. (At first glance one might expect the module to be smaller, but the Class II triangle isn't equilateral.)

The other Class II system we shall find useful is called Method 3 in Clinton's "Geodesic Math" article, where we find it credited to Duncan Stuart. It is one of the most ingenious achievements in the short history of geodesics, permitting as it does an absolute minimum member-length inventory — 4 for 4ν, 12 for 12ν, and so on. We pay for its economy by accepting less symmetrical breakdown triangles. Diagram 10.11 shows the principle; the horizontal members cross the Class II triangle's median at equally spaced intervals, and the bottom edge-point divisions are directly below the corresponding side-edge divisions (broken lines are at 90° angles with bottom edge). It now turns out (Diagram 10.12) that each side-edge division will propagate itself by zigzagging all the way down the main triangle, and each horizontal division will recur as a set of parallel reflections. The figure shows members of eight lengths for 8ν. The chord factors show the asymmetries we accept; all members shorten as we move downward and outward.

Still, the saving in different member lengths is remarkable. This method requires 12 lengths for 12ν. A Class II Method 1 of the same frequency requires thirty-two, and a Class I Method 1 requires forty. It is not surprising that most of the largest geodesic structures hitherto have used Class II Method 3 components.

Of course we realize these savings only with spheres. Once the member length commences to vary with the radius, as in the case of ellipsoids, Method 3 loses all advantage over Method 1. For small Class II spheres, Method 1 gives visibly smoother arcs. But if you need to design a very large geodesic sphere and have any concern at all for economy of fabrication, Class II Method 3 is the place to start.

Which Class?

We may now review the characteristics we have noted in passing.

Class I keeps component variations within narrower bounds, but

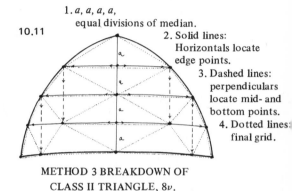

10.11

1. *a, a, a, a,* equal divisions of median.
2. Solid lines: Horizontals locate edge points.
3. Dashed lines: perpendiculars locate mid- and bottom points.
4. Dotted lines: final grid.

METHOD 3 BREAKDOWN OF
CLASS II TRIANGLE, 8ν.

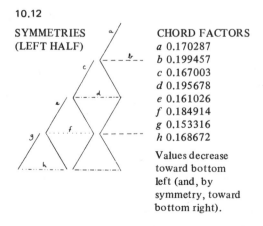

10.12

SYMMETRIES
(LEFT HALF)

CHORD FACTORS
a 0.170287
b 0.199457
c 0.167003
d 0.195678
e 0.161026
f 0.184914
g 0.153316
h 0.168672

Values decrease toward bottom left (and, by symmetry, toward bottom right).

does so at the cost of employing a larger inventory of different strut lengths (how much larger depends on the frequency, and increases rapidly). It permits odd as well as even frequencies, a greater advantage for simple systems than for complex. (Dropping from 4ν to 3ν permits notable economy; the saving of 11ν over 12ν is quite unimportant.) Class I icosahedra have uninterrupted equators at all *even* frequencies, and at all frequencies numerous undulant lesser circles which we can straighten out by a method to be described later and use as truncation planes.

Class II yields even frequencies only. Used with the icosahedron (though not the octahedron or tetrahedron) it introduces some truncation problems, since any equator breaks through some of the subdividing triangles. This means figuring special members to close the gaps. The lengths of the several components vary over a wider range than Class I presents. On the other hand, Class II has economic advantages. It yields somewhat larger triangles for the same frequency than does Class I, somewhat fewer components, and somewhat fewer *different* components. In Method 3, as we have seen, the shrinkage of the list of different components can be spectacular. And we shall see in Chapter 16 its special advantages for diamond-pattern and space-frame applications.

Which Frequency?

The smaller the polyhedron face we are subdividing, the lower the frequency we shall need for acceptable sphericity. And the larger the structure we intend to build, the higher the frequency we shall need, to keep component lengths reasonably short. High frequencies not only use a great many struts, they use a great many different strut lengths, a fact that is less apparent to the eye when the high frequency subdivides an icosa face than when we are dealing with an octa or, worse still, a tetra.

In general, to minimize component inventory, we use the lowest frequency we can. How long a component we can tolerate is a function of its slenderness ratio (length divided by diameter). Fuller has suggested 24/1 as a slenderness ratio for wood, 30/1 for metal or prestressed concrete.

Finally, very high frequencies bring each vertex so near to flatness as to be dangerous unless virtual thickness is imparted to the shell by trussing (see Chapter 16). "Very high," for the icosa, means anything over 12ν with careful workmanship, or over, say, 8ν when tolerances are less than exacting.

11. Choosing a Polyhedron

If our requirements are sufficiently modest, we need not start from a polyhedron at all. Standing at the center of the system we propose to design, we can surround ourselves with as many points as we like, obtained by writing coordinates by hunch, or even at random. We can join each point to its two nearest neighbors to obtain a triangulated network, and use Equation 9.2 to calculate the lengths of these connections. This gives us a table of chord factors. If we then choose a radius and let the chord factors determine the lengths of members, we can build a spherical environment which meets two important criteria. (1) It *is* spherical; every vertex point is at the same distance r from the center. This fact is guaranteed by Equation 9.2, which treats ϕ and θ as spherical coordinates. (2) It is omnitriangulated, hence self-sustaining. We ensured this by our method of joining the points.

But it has grave deficiencies. (1) It is unlikely, except by chance, to possess any symmetries whatever, save that of the spherical envelope (a constant radius). This means that every single member length is probably unique, which is wildly ineffi-cient. It also means that numerous short members are quite likely to be clustered in zones where they make no structural sense, and that elsewhere long unassisted spans are likely to bound large triangular openings. It means that if we want a dome instead of a sphere, no truncation plane exists, nor any logical place to go about inserting one. And it means that the structure looks as unpleasingly random as the design process was.

Moreover (2), it is very unlikely indeed that any sequence of vertices will be aligned along a great circle. *Any two* of them will be, because an equator can be made to pass through any two points on a globe. But that is not what we want to achieve. The geodesic dome, as we have seen, is supported by tension, and the tension networks are most economical when their strands run for considerable distances without changing direction. This means aligning sequences of vertices along great-circle arcs.

Hence the polyhedra, which we described in Chapter 8 as devices for finding symmetry systems.

There are only three such systems to be found: tetrahedral, octahedral, icosahedral. All other regular or semiregular polyhedra present variations on these themes; thus the icosidodecahedron we encountered in Chapter 3 has icosahedral symmetry, and the cuboctahedron has octahedral. It is possible to execute a geodesic breakdown on any polyhedron, by triangulating its nontriangular faces before commencing, and a few polyhedra have special advantages. Thus A. J. Pugh has pointed out that since truncated octahedra cluster naturally, they make a natural starting point for domical clusters (*Polyhedra,* University of California Press, 1976). To isolate first principles, though, we shall concentrate on the three prime symmetry systems, tetrahedral, octahedral, icosahedra.

We commence by setting the tetrahedron aside, for use only in a few very special situations we shall look at much later. Its overwhelming disadvantage is inefficiency. Having only four face triangles, it uses up so much of the sphere's surface before repeating anything that to obtain an average chord factor of about 0.33 we must go all the way to 8ν subdivision (Class I). We then confront twenty different member lengths, varying over a range of 4:1. (By contrast, a 4ν icosa yields a slightly tighter mesh — average chord factor about 0.30 — using only six different lengths which vary over a range of 1.25:1.) A Class II attack on the tetrahedron gives more attractive results and underlies most of its usefulness. It remains a special-case polyhedron merely.

For most purposes, then, we have the icosahedron and the octahedron to choose from.

Useful characteristics of the octahedron include:

- a natural equator at all frequencies;
- vertical slicing line at all frequencies, permitting half a dome to be joined to a rectangular structure;

- simple generation of nonspherical contours;
- provision for square openings: either one at the zenith or, by a simple rotation, three surrounding the zenith partway down;
- relatively simple computations.

Useful characteristics of the icosahedron include:

- lower frequency for comparable average chord factor (a 3v icosa yields sphere subdivisions comparable to a 5v octa);
- angles uniformly averaged among nearly equilateral triangles (octa has right-angled triangles at its principal vertices);
- less variation in the lengths of members for comparable average chord factor (1.18:1 for 3v icosa, 1.5:1 for 5v octa);
- curvature more uniformly distributed (principal octa vertices are nearly flat; hence they are potential weak areas);
- generally superior appearance.

Some of the icosa's advantages disappear once we contemplate nonspherical contours. Since member lengths of elliptical or similar structures must vary continuously from the 0° zenith down to a 90° equator, the icosa loses most of the economy it derives from its generally lower frequency of subdivision. In addition, it requires us to subdivide two different triangles differently en route to the equator, while the octa comprehends this span in a single triangular face.

A minor disadvantage of icosas — lack of a natural equator — disappears at even frequencies, though only with Class I breakdowns. The octa can always be rotated to yield an equator, whether the subdivision is Class I or Class II.

As for the relative ease of computation afforded by the octahedron, this advantage is peripheral; computations, however troublesome, need only be performed once.

There seems to be no doubt that a *wooden* dome is easier to build from octahedral geometries; plywood panels, for instance, fit more easily at square than at pentagonal apices.

In *Shelter* (Bolinas, Ca., Shelter Publications: 1973) Bill Woods of Dyna Domes, whose user-built wooden structures use octahedral geometries and are reportedly trouble-free, is quoted as saying flatly that the icosahedron is "no good." No reasons are given, but Woods presumably had in mind the octa's adaptability to hammer-and-nail technology. Doors and windows, for instance, are easier to fit when squares, not pentagons, are available at the principal vertices.

Other Criteria

We shall see in Chapter 14 that rotating an octahedron until the center of a face is uppermost places three square windows in symmetrical array around the zenith and about 50° down from it. Should we want this trio of skylights, only the octahedron will provide it, and only in a Class I breakdown.

Similarly, the Class II tetrahedron (Method 1), truncated on one of the great circles it makes available, affords at ground level four pairs of large, virtually square openings, each with a single diagonal brace, removal of which leaves a natural door (try an $8v$ model). The tetrahedron's drawbacks, notably its extreme variation in member length, might well be worth coping with should we want a structure with large doors in pairs at the four compass points.

As these examples indicate, some of the inherent difficulties of designing useful geodesic structures can be minimized by choosing the right polyhedron and turning it right way up. Truncation planes will often seem uneven, but in Chapter 14 we shall learn how easy it is to deal with that.

So there may well be practical reasons, deriving from a structure's intended use, for selecting one polyhedron rather than another.

Or there may be aesthetic reasons. A geodesic framing system is normally visible, and we should give some thought to what it looks like. Near-equality of member lengths makes icosahedral systems especially graceful, and functions of the celebrated Golden Section τ ($= [1 + \sqrt{5}]/2$) pervade the proportions of the icosa itself and no doubt help account for its visual appeal. At the other extreme, a designer in quest of bold effects might do worse than investigate the Class II Method 3 tetra, a system of elongated diamonds with incised dihedral angles (again, try an $8v$ model).

12. Using the Tables

We now come to the actual coordinates and their use. Though a glance at the tables in Chapters 18-23 may suggest otherwise, the process by which they are obtained is almost ridiculously simple. Each ϕ, θ pair is generated from a set of 3 integers whose sum is always the frequency of the breakdown, and for the octahedron, especially, the process of generation is so simple we seem closer to numerology than to engineering design. The following paragraphs explain how it is done for the reader who may some day want coordinates for some special frequency. The reader who simply wants to use the tables provided may skip to the next section of this chapter.

Diagram 12.1 shows how we get our sets of three integers. A diagram of one subdivided face is first numbered in the orderly fashion shown in the first drawing; this gives us the first integer of each vertex. The pattern is then rotated 60° clockwise to obtain the second integer, and 60° clockwise again to obtain the third.

Each vertex has now a unique three-integer label, and if we have made no mistakes, the sum of the integers at any vertex will always equal the frequency. Call these integers x, y, z.

Now for the *octahedron*,

$$\tan \phi = x/y \qquad \text{[Eq. 12.1]}$$

$$\text{Class I} \quad \tan \theta = \sqrt{x^2 + y^2}/z \qquad \text{[Eq. 12.2]}$$

$$\text{Class II} \quad \tan \theta = \sqrt{2(x^2 + y^2)}/z \quad \text{[Eq. 12.3]}$$

For the *icosahedron*, we obtain x, y, z in exactly the same way, but before using the equations we convert x, y, z into x_1, y_1, z_1, as follows:

12.1

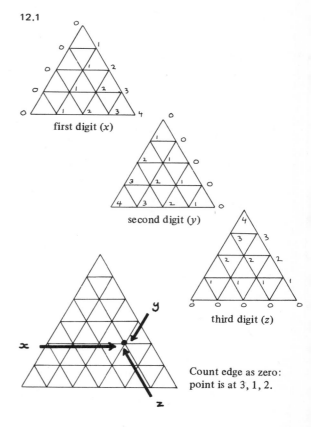

first digit (x)

second digit (y)

third digit (z)

Count edge as zero: point is at 3, 1, 2.

$$x_1 = x \sin 72° \qquad \text{[Eq. 12.4]}$$

$$y_1 = y + x \cos 72° \qquad \text{[Eq. 12.5]}$$

$$z_1 = f/2 + z/\tau \qquad \text{[Eq. 12.6]}$$

Here τ is the famous Golden Proportion, $(1 + \sqrt{5})/2$, or 1.61803399, or $2 \cos 36°$ (this last form is the handiest to enter on the calculator). The icosa coordinates are:

$$\phi = \arctan (x_1/y_1) \qquad \text{[Eq. 12.7]}$$

$$\text{Class I } \theta = \arctan \sqrt{x_1{}^2 + y_1{}^2}/z_1 \qquad \text{[Eq. 12.8]}$$

$$\text{Class II } \theta = \arctan \sqrt{x_1{}^2 + y_1{}^2}/\cos 36 \, z_1$$

$$\text{[Eq. 12.9]}$$

For the *tetrahedron*, we convert x, y, z into x_2, y_2, z_2, as follows:

$$x_2 = \sqrt{3}x \qquad \text{[Eq. 12.10]}$$

$$y_2 = 2y - x \qquad \text{[Eq. 12.11]}$$

$$z_2 = (3z - x - y)/\sqrt{2} \qquad \text{[Eq. 12.12]}$$

The tetra coordinates are:

$$\phi = \arctan (x_2/y_2) \qquad \text{[Eq. 12.13]}$$

$$\text{Class I } \theta = \arctan \sqrt{x_2{}^2 + y_2{}^2}/z_2 \qquad \text{[Eq. 12.14]}$$

$$\text{Class II } \theta = \arctan \sqrt{x_2{}^2 + y_2{}^2}/2z_2 \qquad \text{[Eq. 12.15]}$$

This system gives us Method 1 coordinates. For Class II Method 3 (inferior symmetry, but maximum component economy) we require another system, which is explained in Appendix 3.

Inspection of the octa and tetra equations discloses the useful fact that all coordinates for these polyhedra are either rational fractions or the roots of rational fractions. Thus 2,3,3 in an 8ν octa has the coordinates

$$\phi = \arctan 2/3 \qquad (33.6900675°)$$

$$\theta = \arctan \sqrt{13}/3 \quad (50.2378408°)$$

Both values are given in the tables, but the former is quicker to enter on the calculator.

We shall now see how to use coordinates to design a geodesic structure.

Planning the Work: Class I

Sketch an equilateral triangle (Diagram 12.2), subdivide the edges into *n* parts for *n* frequency, and join the points of subdivision with a three-way grid.

Now start with the top vertex (0,0) and number all the crossing points, as shown. For frequency *n* the last point on the bottom row will be *n, n.* Every point now has a two-digit marker.

Now draw the three medians of the face triangle (vertex to midpoint of opposite side: the broken lines in Diagram 12.2). You have outlined six right-angled triangles, each of which contains the whole symmetry system of the polyhedron. Concentrate on the top left one, which we'll call "the" symmetry triangle. *If the structure we are planning is spherical, we need chord factors only for the breakdown edges that lie partly or wholly inside the symmetry triangle.* For the 6*v* example shown, this means only the twelve edges drawn with heavy lines: twelve out of sixty-three. For a sphere the only coordinates we need are the endpoints of those twelve edges: a total of ten coordinates. The rest of the system repeats symmetrically; thus 0,0/1,0 corresponds to 5,0/6,0; 6,0/6,1; 6,5/6,6; 5,5/6,6; 0,0/1,1: six occurrences of the one length. Anything that cuts an edge of the symmetry triangle occurs only three times per face: thus, 3,1/3,2 is repeated at 4,1/5,2 and 4,3/5,3.

All this is true of spheres. If we plan something elliptical we need more lengths, because the radius changes constantly as we move down from the zenith (0,0), and the top half of a face won't duplicate the bottom half. We need coordinates as far down as $\theta = 90°$, where the bend of the elliptical contour reverses itself. An octa or tetra face goes down that far; an icosa face doesn't, and the icosa tables provide for a second face, drawn under the first and numbered as shown in Diagram 12.3. (For more details see Chapter 13.) Because of the steadily altering radii the symmetry

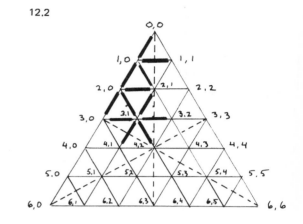

12.2

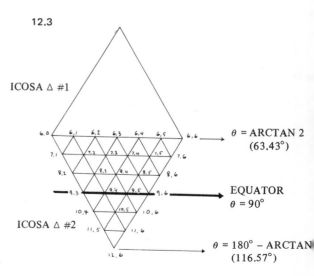

12.3

ICOSA △ #1

ICOSA △ #2

θ = ARCTAN 2
(63.43°)

EQUATOR
θ = 90°

θ = 180° − ARCTAN
(116.57°)

triangles no longer contain all the different lengths, but we still have left- and right-hand mirror symmetry to cut down our work.

We now make a list of the coordinates we actually require. To keep the example simple, we'll stick to the sphere.

Each face has actually six symmetry triangles, and we'd get the same chord factors whichever we investigated. But we use the top left one because the numbers to be manipulated are handier if we stay to the left, where many ϕ values are zero, and toward the top, where 0,0 occurs and θ values are smallest.

For a 6ν sphere the worksheet would look like this:

				ϕ	θ
0,0/1,0					
1,0/1,1			0,0		
1,0/2,0			1,0		
1,0/2,1			1,1		
2,0/2,1	chord	coordinates	2,0		
2,0/3,0	factors	required	2,1		
2,0/3,1	required		3,0		
2,1/3,1			3,1		
3,0/3,1			3,2		
3,1/3,2			4,1		
3,1/4,1			4,2		
3,1/4,2					

Now we know which coordinates to extract from the table. Assume we want an icosahedral dome. We turn to "Icosahedron Class I Coordinates: Frequencies 6,3" (Chapter 20). Going down the "6ν" column to the vertices we have just listed, we copy the coordinates as we find them. The list looks like this:

	ϕ	θ
0,0	0	0
1,0	0	9.3247035
1,1	72	9.3247035
2,0	0	20.0767513
2,1	36	16.4722107
3,0	0	31.7174744
3,1	22.3861776	27.2237351

3,2	49.6138225	27.2237351
4,1	16.0353713	39.1034177
4,2	36	37.3773682

Now we are ready for chord factors. Since our structure is to be spherical we use Equation 9.2 and insert pairs of ϕ, θ values from this table. Thus for the distance of 0,0/1,1 we use 0 for the values of ϕ_1 ϕ_2 and θ_1, and 9.3247035 for θ_2. The chord factor turns out to be 0.162567. The reader should now practice with the calculator and the equation until he gets the following results. Once your routine is debugged you can work with great confidence.

0,0/1,0	0.162567	2,0/3,1	0.198013
1,0/1,1	0.190477	2,1/3,1	0.205908
1,0/2,0	0.187383	3,0/3,1	0.205908
1,0/2,1	0.181908	3,1/3,2	0.215354
2,0/2,1	0.202820	3,1/4,1	0.215354
2,0/3,0	0.202820	3,1/4,2	0.216628

We see that we have more symmetries than we had been led to expect: there are three identical pairs, hence only nine different component lengths overall. With experience such duplications can be anticipated. But it is always safer to invest a couple of minutes figuring a superfluous value or two than to waste time and building materials on a false assumption.

A way of checking the work is to list the coordinates inside a different symmetry triangle entirely and calculate the values for these. Though we are working with completely different numbers, the chord factors should be identical.

Note that the longest members radiate from the point (4,2) where the symmetry triangles meet, and the shortest (0,0/1,0) from the corners of the face triangle. This is generally true of geodesic spheres, and gives us a quick test for improbable results. These two lengths are the extremes, and the rest should fall smoothly between them.

Accuracy

We have given six-figure chord factors, and it is interesting to see what sort of accuracy this would yield in practice. We might envisage a hemispherical dome 100 feet in diameter (50-foot radius). This would entail strut lengths of the order of 10 feet, bigger than anyone is likely to go with 6ν. The chord factor nearest the mean value (0.196) is 0.198013, and multiplying this by 50 feet we obtain a strut length for these particular members of 118.80780 inches. If we change the chord factor by +1 in the sixth place, we get 118.8084 inches, an error of 6 ten-thousandths of an inch. That is as much trouble as we are likely to get into by not using seven-place chord factors. Our coordinate tables are given to seven places (various rounding-off errors make the eighth place probable but not certain), so six places seems far enough to take chord factors.

Class II

The design of a Class II structure goes in the same way, with a few variations. The triangle we draw and subdivide is not the polyhedron face but its Class II triangle, as shown in Diagram 10.10. It straddles the edge between two polyhedron faces, and its bottom edge joins their centroids. For n-frequency division of the polyhedron, we make an $n/2$ subdivision of its Class II triangle, and our module of symmetry is the entire left-hand half of this triangle. Diagram 12.4 shows the idea. For a 6ν Class II subdivision of the icosa face, we make a 3ν subdivision of its Class II triangle. Chord factors for the left half of this are duplicated in the right half, and the entire Class II triangle is duplicated by mirror symetry below itself, bringing us to the end of an icosa edge, which we find we have thus divided into 6. So much for Class II spheres.

For the ellipse things are a little more complicated. To get down to the equator we need five symmetry triangles, as shown in Diagram 12.5 for an 8ν example. Note carefully how the numbering is done; the sequence 0,0; 1,0; 2,0; and so on proceeds down the left-hand edge, and such sequences as 4,0; 4,1; 4,2 follow the *upward*-slanting lines.

12.4

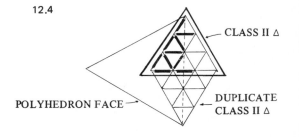

CLASS II △

POLYHEDRON FACE

DUPLICATE CLASS II △

The equator ($\theta = 90°$) always passes through the midpoint of the edge the two bottom Class II triangles share (10,2 in the example) and passes a little below the edge point at the left, a little above the edge point at the right. A glance at the 8ν icosa Class II Method 1 table confirms this; 1,0 has a θ of 90, but the corresponding values for 9,0 and 11,4 are 88.95519820 and 91.04480176. So an equatorial cutoff just misses alternate vertices;

In Chapter 10 Class II Method 3 was presented as a system for maximizing the economy of different component lengths, at the cost of a slight loss in triangle regularity. Since this virtue applies only to spheres, the designer will be asking of this method none but spherical chord factors. So a complete set of these is provided in Chapter 23. They are ready for use without reference to the coordinates, which are presented only for their usefulness in two contingencies: the insertion of truncation planes (Chapter 14) and the design of space-frame structures (Chapter 16).

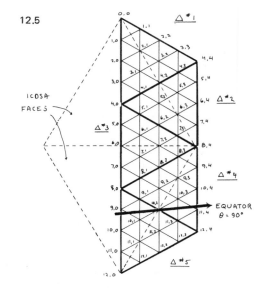

Derivation of the Coordinates

Why the spherical coordinates are generated from triads of small integers we may learn from Diagram 12.6, which shows one face ABC of an octahedron with a 3ν breakdown outlined upon it. The breakdown is projected downward onto the horizontal plane OBC, every point on the projection lying directly beneath its original.

12.6

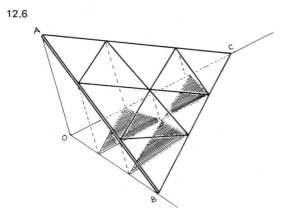

If we now look at the projection plane OBC by itself (Diagram 12.7) we see that it is a $45°$ right-angled triangle, OB and OC indistinguishable from the Cartesian x and y axes. Thus all points on the projected breakdown have simple integral x,y coordinates, and so do the points on the octa face directly above them (p is at 1,1).

12.7

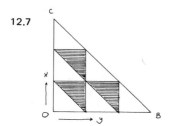

Diagram 12.8, a side view, shows how the $45°$ slope of the octa face translates integral distances ($O\text{-}p_1$) from the origin into integral verticals ($p_1\text{-}p$). Thus the z coordinates of the breakdown vertices are also integral.

12.8

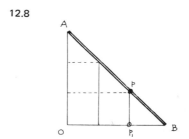

Turning now to Diagram 12.9, which is Diagram 12.6 redrawn with some details omitted, we see that the ϕ of point p is simply $\arctan x/y$. Its θ is $90°\text{-}$(angle pOp_1). Thus $\tan\theta$ is Op_1/z. Pythagoras tells us that $Op_1 = \sqrt{x^2 + y^2}$, hence $\theta = \arctan\sqrt{x^2 + y^2}/z$.

12.9

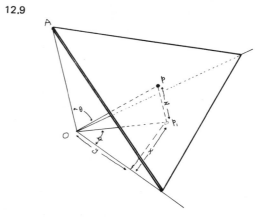

Before leaving the octahedron, we may recall A. J. Pugh's observation that its truncated form, being a space-filling polyhedron, makes a sensible configuration to start from if we envisage clustered domes (*Polyhedra*, University of California Press, 1976). The coordinates of truncated octahedral breakdowns are equally easy to write. Diagram 12.10 shows why, to get a frequency f (here 3ν) we start from an octahedral subdivision of $3f$ (here 9ν). We divide the octa face edges into thirds and fold the top corner back until it is parallel with the projection plane; it will become a quarter of the square face, our triangle 1. Its x and y values remain as they were, but its parallelism with the projection plane makes its z retain everywhere the value of

12.10

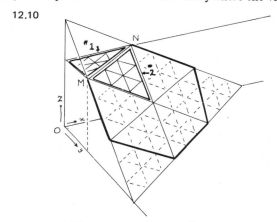

z along line MN, that is, $2f$ (here = 6). Triangle 2 is a portion of the octa face, unchanged from its former orientation, so we can evaluate its points as if they were details in a $3f$ octa subdivision. Using our rule for writing the x,y,z of single points, the reader can see that point 1 in Diagram 12.11 is at 3,3,3; point 2 at 3,2,4; point 4 at 2,2,5; and so forth. And remembering to keep $z = 2n$, he can identify

12.11

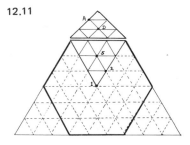

point A as 0,1,6; point D as 1,1,6; and so on. Putting these values into the octahedron equations (Equation 12.1 and Equation 12.2) we can quickly derive the coordinates of all the points we need; thus A is at 0, 9.4623222.

Diagram 12.12 shows what happens when we project an icosa face $O_1A_1C_1$ onto the plane where the Cartesian axes OB, OA are located. The $72°$ angle at the icosa apex makes the projection lie skew. In Diagram 12.13 we see the projection AOC nestled between equal axes AO, OB. The point P is midway along AC; we choose this special case to keep the diagram simple, but any point on the icosa face would do. Q is similarly located at the midpoint of AB. Its coordinates are x,y, and we want to find the coordinates $x_1 y_1$ of P. If $OC = OB = 1$, it is easy to see that $DC = \sin 72°$; $DC/OB = \sin 72°/1$; and $x_1 = x \sin 72°$, as in Equation 12.4. Similarly, $y_1 = y + PQ$. But $PQ/x_1 = \tan 18°$, so $PQ = (x \sin 72°)(\tan 18°) = x \cos 72°$. Thus $y_1 = y + x \cos 72°$, as in Equation 12.5.

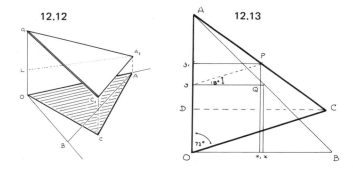

12.12   **12.13**

For z_1 we turn to Diagram 12.14, which is the rear wall of Diagram 12.12, with the same lettering. O_1A_1 is the icosa edge, O_1O is the z axis with O at the center of the polyhedron. We shall take for granted two facts inherent in the icosa's governing ratios: that in the triangle O_1A_1L, tan $A_1 = 1/\tau$, and that in the rectangle $LOAA_1$, $OA = 2LO$. (The reader with an appetite for airtight proofs is welcome to construct them.)

We now divide the icosa edge into f equal parts (here 3), and by dropping perpendiculars divide LA_1 and OA likewise. We remember that $f = x + y + z$, and since OA is the

y-axis, $x = 0$ and $z = f - y$ (again we are using a special case for simplicity). The z_1 into which we propose to convert this consists of two parts, the part in the triangle O_1A_1L, the part in the rectangle $LOAA_1$. The former is $(f - y)/\tan A_1$, that is, z/τ. The latter is equivalent to LO, which is half OA; but $OA = f$. Hence $z_1 = f/2 + z/\tau$, as in Equation 12.6.

The tetrahedron is analyzed in the same manner, and the reader who wants proofs of Equation 12.10, Equation 12.11, and Equation 12.12 should have no difficulty working them out.

12.14

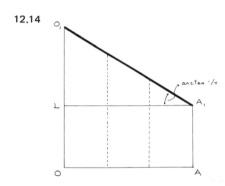

arctan $1/\tau$

13. Ellipses and Superellipses

So far we have acquired nothing, except some understanding, that couldn't be contained in a comprehensive table of chord factors. The real interest of the spherical coordinates, making them worth the labor of computing them, is the way they facilitate geo-desicizing nonspherical envelopes. The obvious profile to start with is the ellipse, and the most compact way of tackling the ellipse is to think of it as a member of a large family of shapes with a common mathematical description.

We may look at the general equation first, and then move to simple cases by taking things out of it:

$$\frac{x^k}{a^k} + \frac{y^m}{b^m} + \frac{z^n}{c^n} = 1.$$

Here x, y, and z are ordinary Cartesian coordinates, at $90°$ angles to one another; k, m, n, a, b, c are any positive quantities you like. By varying them we can describe a surprising array of symmetrical shapes: a brick, a modified egg, a sphere, an octahedron (stretched or regular), a pair of crossed lines, a circle, any ellipse at all, not to mention various cigar shapes and squares with rounded corners. We may start in two dimensions by letting $z = 0$, which suppresses the third fraction entirely. Then if a and b are both 1, and k and m both 2, we get

$$x^2 + y^2 = 1,$$

which describes a circle centered at 0,0.

The radius of the circle, of course, doesn't change. To get a changing radius we insert the terms a and b. It is easiest to let $a = 1$ and take up all the variation in b. This gives us

$$x^2 + \frac{y^2}{b^2} = 1,$$

which describes an ellipse. We see at once that if $b = 1$ this equation is no different from that of the circle. Obviously b determines how the radius varies as we travel round the curve. It is, in fact, the ratio (see Diagram 13.1) of the long and short axes, and to help ourselves remember this we may give it a name, E, for *expansion*.

So the radius of the ellipse will vary between E and 1, and an ellipse with an expansion of, say, 1.5 will be 1.5 times as long as it is broad. If this curve describes the profile of a dome with equatorial truncation, then the dome will have a radius of 1 at ground level and a radius of E at the zenith. It is handy to let the floor radius = 1; then if $E > 1$ the dome will be tall for its girth, and if $E < 1$ the dome will be flattened. We can thus increase or diminish the headroom without altering the area of the circular floor.

To design the dome we need to know exactly how the radius varies. We know that $x^2 + (y^2/E^2) = 1$, and from Diagram 13.2 we see that $x = r \sin \theta$ and $y = r \cos \theta$. Inserting these values for x and y and rearranging, we obtain our first useful expression:

$$r = \sqrt{E^2/(E^2 \sin^2 \theta + \cos^2 \theta)} \quad \text{[Eq. 13.1]}$$

In Chapter 9 we calculated a very simple (2ν octa) dome using Equation 9.1, which gives us chord factors between two points of different radius. E was 1.5, and we made a θ:r table in which one value had to be taken on trust. (It is now evident how it was arrived at.) For more ambitious cases we proceed in exactly the same way:

1. from the coordinate table, list the values of θ that occur between the zenith and the truncation plane;
   2. using Equation 13.1, obtain a value of r for each θ; and
   3. use Equation 9.1 to calculate the chord factors, being careful to use the proper pair of rs for each pair of θs.

13.1

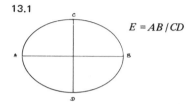

$E = AB/CD$

13.2

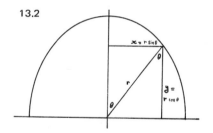

Octahedral Ellipsoids

As usual, the octahedron is the easiest to handle. Its natural equatorial truncation keeps us out of trouble. Sliced at the equator, it yields a beehive shape with a circular floor. By slicing it down the long axis we obtain an elliptical floor, with a semi-elliptical profile as viewed from the side and a semicircular profile as viewed from the end.

Such figures, of which one profile is circular, are called *ellipsoids of revolution,* because we can imagine their being generated by twirling an ellipse.

EXAMPLE 1. Diagram 13.3 shows the coordinate diagram for a 4v octa. For E = 1.5, *the θ values and radii are:*

13.3

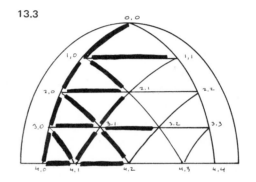

θ	r
0	1.5
18.4349488	1.4142136
35.2643897	1.2602521
45.0	1.1766968
65.0051575	1.0497813
71.5650512	1.0289915
90.0	1.0

Chord factor 0,0/1,0 is obtained with Equation 9.1, using the radii of 0,0 and 1,0 (1.5 and 1.4142136, respectively). It is 0.474423. Fifteen more chord factors are calculated in this way, accounting for every length in the left half of the diagram. Mirror symmetry takes care of the rest.

Chord Factors

0,0/1,0	0.474423	2,1/3,1	0.696631
1,0/1,1	0.632456	3,1/3,2	0.606092
1,0/2,0	0.638578	3,0/4,0	0.326266
1,0/2,1	0.605791	3,0/4,1	0.454575
2,0/2,1	0.635872	4,0/4,1	0.320364
2,0/3,0	0.526758	3,1/4,1	0.452409
2,0/3,1	0.589150	4,1/4,2	0.459506
3,0/3,1	0.456607	3,1/4,2	0.532697

Icosahedral Ellipsoids

The five-way symmetries of the icosahedron make for especially beautiful ellipsoidal patterns. This important advantage is its principal one. As soon as we depart from spherical profiles, we lose a virtue we have come to expect of the icosa, a relatively small inventory of different strut lengths. Unless we are going to settle for ridiculously shallow truncations, we must cope with two different face triangles. This is clear from Diagram 13.4, where we see the icosa dissected into (1) Fuller's "icosacap," the set of five triangles surrounding the zenith, and (2) the set of ten triangles around the waist. It is clear that we must enter this second set to find the equator, and that the two sets will be stretched or squashed in different proportions as the radius swings down. So the icosa Class I tables provide coordinates for a second triangle. In the same way, the icosa Class II tables cover five Class II triangles: Diagram 13.5 shows how they are located on the icosa.

   We shall work an example from each class, and to keep the work brief we shall stick to 2*v*.

   EXAMPLE 2: 2v icosa Class II. Diagram 13.6 shows the vertex diagram for the five Class II triangles. The coordinates are quickly listed from the table, and the radii calculated:

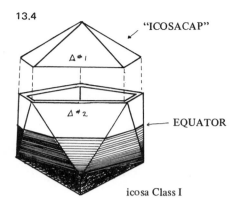

13.4

"ICOSACAP"

Δ #1

Δ #2

EQUATOR

icosa Class I

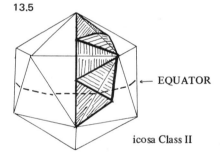

13.5

EQUATOR

icosa Class II

13.6

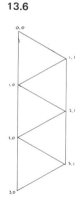

	ϕ	θ	r
0,1	0	0	1.5
1,0	0	37.3773681	1.2411302
1,1	72	37.3773681	1.2411302
2,0	0	79.1876830	1.0099209
2,1	36	63.4349488	1.0606601
3,0	0	116.5650512	1.0606601
3,1	36	100.8123170	1.0099209

The symmetries above and below the 90° line are evident from the table of radii above: 2,0 corresponds with 3,1 and 2,1 with 3,0. To see how this works out in distances, we shall this time list all the chord factors:

Chord Factors

0,0/1,0	0.911918	2,0/2,1	0.665207
0,0/1,1	0.911918	2,1/3,1	0.665207
1,0/1,1	0.885725	2,0/3,1	0.720724
1,0/2,0	0.831758	2,0/3,0	0.665207
1,0/2,1	0.757107	3,0/3,1	0.665207
1,1/2,1	0.757107		

There are eleven distances in the diagram, but only six different ones. The builder would need only six different members.

Since θ = 90 occurs nowhere in the coordinate table, no point on this structure lies on an equator. Indeed, we may wonder how we should go about truncating it. Chapter 14 will tell.

We move on to the second icosa example, a Class I ellipse. This time we shall use a fractional expansion, 1/1.5, to get a saucer rather than a beehive. We shall have fault to find with the first version we come up with, and in learning how to modify it will master a useful principle.

EXAMPLE 3: 2v icosa Class I, squashed. The coordinate layout is shown in Diagram 13.7. The figures look like this:

13.7

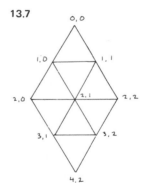

	φ	θ	r (E = 1/1.5)
0,0	0	0	0.66666667
1,0	0	31.7174743	0.72461704
1,1	72	31.7174743	0.72461704
2,0	0	63.4349488	0.89442719
2,1	36	58.2825255	0.86210372
3,1	18	90.0	1.0
3,2	54	90.0	1.0

Though the coordinate table goes past 3,1 and 3,2, we don't need any values beyond 90 (the equator).

We now compute chord factors and get the following:

0,0/1,0	0.384256	2,0/2,1	0.481007
1,0/1,1	0.447838	2,0/3,1	0.527551
1,0/2,0	0.471622	2,1/3,1	0.590178
1,0/2,1	0.507459	3,1/3,2	0.618034

Modifying Coordinates

Study of the chord factors we have just calculated, or better still study of a simple model, will disclose an unwelcome asymmetry: the short members are bunched around the zenith, where they have little to contribute but weight, and the longest members, straddling the most widely separated vertices, are found around the sides where the load is being transmitted to earth. A glance at Diagram 13.8 shows what has happened. Our coordinate system was designed to give reasonably uniform member lengths on a sphere. When a shallow curve intercepts the bundle of sight lines, different struts are differently affected. This is true of course with any noncircular contour, but it is more likely to be troublesome with a shallow ellipse. A beehive may often concentrate its elongations near the zenith with neither aesthetic nor structural harm, since the zenith members have little load to bear.

Our recourse is to modify the coordinates before doing anything else. Instead of projecting the θ array against an expanded envelope, we might in effect allow the expansion of the envelope to pull the zenith coordinates apart. This is like reflecting the array off a slanted mirror whose slope is governed by the expansion (Diagram 13.9). Each θ is replaced by θ_1, obtained from

$$\theta_1 = \arctan(\tan\theta/E)$$

Diagram 13.10 shows a typical result. The stretching of chords now tends to parallel the long axis of the ellipse — the reverse of the situation in Diagram 13.8 — and the zenith of a shallow structure is no longer congested with short members. Calculation would also show that the overall variation in strut length is reduced. This procedure gives chord factors identical with those of the elliptical structures in *Domebook 2*. We may nickname it the E Correction.

But there is a simple way to do still better. Replace each θ by θ_1, obtained from

$$\theta_1 = \arctan(\tan\theta/\sqrt{E}) \qquad \text{[Eq. 13.2]}$$

We may call this the Root E Correction. It distributes the strut lengths very much as the parent sphere did, the longest members part-way down, shorter ones toward zenith and base. And the

13.8

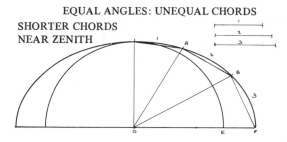

EQUAL ANGLES: UNEQUAL CHORDS
SHORTER CHORDS NEAR ZENITH

13.9

OE/OF = EXPANSION
OA, OB: OLD SIGHTLINES
OC, OD: NEW SIGHTLINES

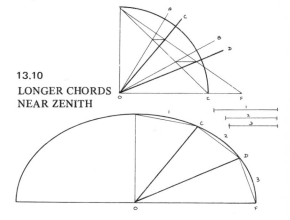

13.10
LONGER CHORDS NEAR ZENITH

overall variation of strut lengths is held within close bounds —
probably the closest we can easily obtain. In fact the Root E
Correction may be recommended as a routine procedure for all
elliptical geodesics. For the case we have been discussing it gives
these chord factors:

0,0/1,0	.456149	2,0/2,1	.515500
1,0/1,1	.529609	2,0/3,1	.475770
1,0/2,0	.472080	2,1/3,1	.530104
1,0/2,1	.542841	3,1/3,2	.618034

As an exercise, the reader may refigure the 2ν Class II icosa
beehive we figured previously, and observe the improvement that
results.

Superspheroids

At the beginning of this chapter we displayed a general equation
from which we obtained the equation of the circle by assigning
appropriate values. To the two exponents k and m we assigned the
value 2, and we kept 2 for an exponent when we went on to the
ellipse. But there is no reason any exponent needs to be 2. It can
be anything from zero to infinity. If it is zero, the ellipse or circle
equation yields not a curve but a pair of crossed lines. If it is
infinity, we obtain a rectangle (or, if $E = 1$, a square). Some years
ago the Danish polymath Piet Hein proposed mediating between
the circle and the square by making the exponent 2.5. If we
modify Equation 13.1 in this way, remembering to insert a 2.5
root as well as 2.5 powers —

$$r = \sqrt[2.5]{E^{2.5}/E^{2.5} \sin^{2.5} \theta + \cos^{2.5} \theta} \quad [\text{Eq. } 13.3]$$

— we have once again a formula for a varying radius, that of a
supercircle if $E = 1$, and of a *superellipse* if E has some other value,
for instance 1.5.* The supercircle radius formula by itself looks
like this:

*See Martin Gardner's article in *Scientific American* (September 1965):
222-234.

$$r = \sqrt[2.5]{1/(\sin^{2.5}\theta + \cos^{2.5}\theta)} \qquad \text{[Eq. 13.4]}$$

A supercircle (Diagram 13.11) is beginning to develop shoulders. A superellipse (Diagram 13.12) has less taper than a normal ellipse of the same expansion. A dome with either profile retains more of its headroom further from the center than does the nonsuper version. The versatile Peter Calthorpe designed a superspherical 3ν icosa some years ago. The one effort I know of to build one encountered legal and financial troubles unrelated to the design.

The simplest way to use these shapes is to combine them with a circular floor plan as we did the ellipse. We thus obtain the shape generated by twirling a supercircle or a superellipse on its axis: a superspheroid or a superellipsoid of revolution.

The x^y (or y^x) function on the calculator makes 2.5 almost as easy an exponent to use as 2. (We get the 2.5th root by raising x to the power $1/2.5$). Radius tables and then chord factors are obtained exactly as they were earlier in this chapter. The reader should have no difficulty obtaining the following chord factors for a 4ν octa supersphere:

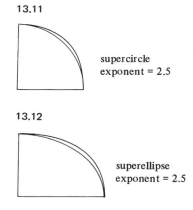

13.11

supercircle
exponent = 2.5

13.12

superellipse
exponent = 2.5

0,0/1,0	0.326072	3,0/3,1	0.454425
1,0/1,1	0.459825	3,0/4,0	0.326072
1,0/2,0	0.484335	3,0/4,1	0.454351
1,0/2,1	0.460526	3,1/4,1	0.449895
2,0/2,1	0.552911	4,0/4,1	0.320364
2,0/3,0	0.484335	4,1/4,2	0.459506
2,0/3,1	0.547799	3,1/4,2	0.529997
2,1/3,1	0.608369	3,1/3,2	0.601483

Two of these are duplicated, so the 4ν octa supersphere requires 14 different components, as against 6 for a normal sphere. Careful study of a model is the only way to decide if it is worth the complication.

Just as 2 is an optional exponent, so is 2.5. The latter has no special magic except that Piet Hein likes the curves it generates. By raising the supersphere exponent to 3, we can retain headroom even further from the center. But another way of putting this is that we are making a large area around the zenith nearly flat (see Diagram 13.13). The spherical excess of the structure is concentrated in a rather narrow zone where roof suddenly becomes wall. Nearly flat areas in geodesic structures are areas of potential trouble, requiring careful hub detailing to retain the integrity of the tension network. Stretching the supercircle into a superellipse expands the problem area, and exponents above 2.5 are not recommended for superellipses.

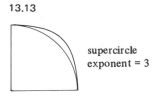

supercircle
exponent = 3

Varying Two Radii

The sphere is a circle of revolution; the supersphere is a supercircle of revolution; the ellipsoids we have considered have been ellipsoids of revolution. This simply means that one cross section is circular, and such domes, unless they are truncated obliquely, will have either a circular floor plan and a special profile, or a special floor plan and one semicircular profile. The structure's native sphericity is expanded (or contracted) in only one direction.

But there is no reason why we cannot vary two radii. This means specifying two separate expansions, E_1 and E_2, one for the floor and the other for the profile. (*Separate* need not mean *different*; both expansions may have the same value, but they will both make their presences felt.) So we could have a supercircular profile, to get headroom, and a superelliptical floor plan, to get walking space. The respective expansions would be 1 and something else, say 1.25. The exponents could be the same, or different. The structure would have no circular cross section at all.

Since the radius of each vertex is now responding to two expansions, and perhaps to two exponents, the arithmetic becomes more labyrinthine. Still, all we need to calculate chord factors is a table of radii as measured from a central point, and we can make one if we keep our heads.

Diagram 13.14 shows the situation. We want the radius from the center of the system to vertex ϕ, θ. We first assume a floor plane passing through the center (though when we are finished nothing will prevent us putting the floor somewhere else). We next want r_1, which is the radius of the floor at the point where the θ meridian through the vertex in question crosses its edge. This

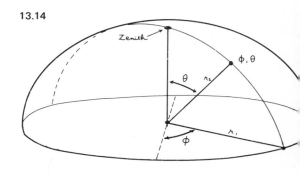

radius is a function of E_1, the expansion of the floor ellipse, and of the ϕ of the vertex in question. Its equation is:

$$r_1 = \left\{ E_1{}^{n_1} / (\cos^{n_1} \phi + E_1{}^{n_1} \sin^{n_1} \phi) \right\}^{1/n}$$

[Eq. 13.5]

We might tabulate r_1 for every pertinent value of ϕ, which means for one-quarter of the dome's perimeter, and use it to obtain r_2, the radius we want for chord factors, as follows:

$$r_2 = \left\{ r_1{}^{n_2} E_2{}^{n_2} / (E_2{}^{n_2} \sin^{n_2} \phi + r_1{}^{n_2} \cos^{n_2} \theta) \right\}^{1/n}$$

[Eq. 13.6]

But we should go to all this trouble only in the unlikely case when we might want to use different values of the exponent n for the two radii. The exponent is, of course, 2 for ellipses and normally 2.5 for supercircles and superellipses, and it seems ill-advised to mix the two in one structure. So we take advantage of the fact that when the two exponents are identical the two equations collapse into one:

$$r_2 = \left\{ (E_1 E_2)^n / (E_2 \sin \theta)^n \, (\cos^n \phi + [E_1 \sin \phi]^n) + (E_1 \cos \theta)^n \right\}^{1/n}$$

[Eq. 13.7]

(The exponent $1/n$, if you've forgotten, means the nth root.)

For high jinks of this kind the octahedron has obvious advantages, since one face triangle takes care of the whole quadrant and brings us clear down to the equatorial truncation line. In Chapter 15 we shall look at its disadvantages, which stem from the way its tendency toward flatness at the square vertices is accentuated by the flat areas peculiar to the superellipse.

The icosa remains elegant and far from impossible. In Chapter 15 we shall develop a fancy application in detail, and a worked example is deferred until then.

A few pages back we found that it was sometimes desirable to improve the symmetry of a non-spherical form by modifying one coordinate; Equation 13.2 gave a convenient method. Now that we are dealing with two expansions the symmetry has more ways of becoming displeasing, and improving it will mean modifying both coordinates. An elegant result is obtained by changing ϕ to ϕ_1:

$$\phi_1 = \arctan\,(\tan\phi/E_1) \qquad \text{[Eq. 13.8]}$$

and θ to θ_1:

$$\theta_1 = \arctan\,\left\{\tan\theta\;(E_1{}^n/\cos^n\phi_1 + E_1{}^n\sin^n\phi_1)^{1/n})/E_2\right\}$$

$$\text{[Eq. 13.9]}$$

These expressions reproduce the effect of projecting points onto a sphere and then expanding it, rather than projecting spherical coordinates onto an already expanded envelope. The designer will have to judge whether they are worth the bother.

They assume, too, that the floor is being expanded along the $\phi = 0°$ axis, an innocuous assumption if we are using the octahedron, whose 0° and 90° axes are indistinguishable. Chapter 15 shows that with the icosahedron there is reason to make a decision, and should we decide to expand its floor along the $\phi = 90°$ axis, we need to alter both equations slightly. In Equation 13.8 the division becomes a multiplication, and in Equation 13.9 "sin" and "cos" must be interchanged.

Free Forms

It is possible to apply the geodesic three-way grid to any surface that can be mathematically described. The description is not always easy, and one can conceive of convex curves that would give a skilled mathematician trouble. One thing that is simple in spherical coordinates, though, is the notation for any surface of revolution whose profile is a convex curve drawn freehand. Should you want a shape like the one in Diagram 13.15, you have only to draw it on graph paper, designate an origin point, and from that point run radii out to the curve, the angle of each radius with the vertical axis being one of the θ values for the breakdown you have chosen. Call the length of the vertical axis $r = 1$, and measure the other radii. Then use those radii, and the ϕ,θ values from the table for your breakdown, in Equation 9.1 to get chord factors. The octahedron's 90° seams make it possible to combine two such profiles, left and right. Keep the roofline convex and avoid extended arcs of very low curvature, unless you are prepared to take the structural measures described in Chapter 16.

13.15

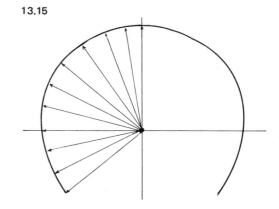

14. Truncations

So far we have been accepting symmetry systems just as they come from the mathematician's repertory. We made certain choices: the rest was law. We chose a polyhedron, we chose a frequency and a breakdown, we chose floor and roofline cross sections with an expansion and an exponent for each; having chosen, we accepted what came out of the calculator. It is now time to contemplate a few interventions, which modify the crystalline symmetry of the result to make it more useful. In particular, we want to be able to slice our complex polyhedron at will and make it sit flat; we need to design *truncation planes* into it.

A few of these are inherent, and always equatorial. The octahedron has natural equators, the inherent great circles, and any breakdown preserves some of them. Complete icosahedral great circles are preserved by any *even-frequency* Class I breakdown. Class II icosa great circles are tantalizing: they are twice interrupted in their circuit around the sphere, and they lie flat only when the sphere is tipped so that an edge, not a vertex, is at the zenith. To get at them will take special handling: we'll come to that.

If we want something other than a hemisphere — in short, a *lesser-circle truncation* — we find that no combination of breakdown and frequency will give it to us. We shall have to tamper.

There are two main ways of tampering. One is to devise a *truncatable breakdown,* which supplies crisscrossing lesser circles throughout the system. Any of these will then serve as a truncation plane. Diagram 14.1 gives a rough idea. In its raw state the 3*v* icosa is girdled by twelve approximate lesser circles, running in pairs like the Tropic of Cancer and the Tropic of Capricorn, parallel to six imaginary equators (the six great circles, which the

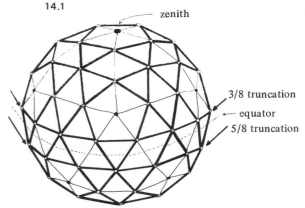

14.1

zenith

3/8 truncation

equator

5/8 truncation

3*v* CLASS I
LESSER-CIRCLE SYSTEMS

breakdown does not incorporate). They are so nearly level that many builders have simply chosen to pretend they were truncation planes, and have shored up the wobble with blocks (see Lloyd Kahn's *Domebook 2* [Bolinas, Ca., Pacific Domes: 1971] p. 24). Truncatable breakdowns level all twelve of these circles off, giving the builder a choice of two genuine truncation planes, at approximately 3/8 and 5/8 of the sphere's volume.

This attractive method leads to a couple of interrelated difficulties. Tampering with the overall geometry can lead to poor distribution of spherical excess; avoiding that can lead to surprisingly wide variations in strut length. Some of the earliest truncatable designs incurred the first of these difficulties: the pentagon vertices were excessively peaked, leaving the rest excessively flat. Other solutions trade asymmetry of spherical excess for asymmetry of member length, chord factors varying as much as 45 percent (vs. about 17 percent for the normal $3v$). Truncatables do not need to be unhandsome: an especially fine $5v$ roofs the religious center at the Edwardsville campus of Southern Illinois University, strut lengths varying in a sequence so elegant it looks more like an aesthetic modulation than an engineer's contrivance (*Architectural Forum*, January-February, 1972:86, fig. 5).

Still, for many purposes the second main solution to the truncation problem is preferable. It amounts to leaving most of the design alone and inserting just one truncation plane where we need it.

Spherical coordinates make this extremely simple. Since *all points with the same θ lie on the same lesser circle*, we have only to assign the same θ to contiguous points and a truncation plane will pass through them.

EXAMPLE 1. Diagram 14.2 shows one face triangle of a 5v octa, Class I, which we propose to truncate along the line 4,0-4,4. Three different θ values appear on this line, 75.9638, 72.4516, 70.5288. To secure a truncation plane we have only to make them all equal and recalculate the chord factors for all struts connected to the points whose location we are changing. We can select one of these values — say, the edge one — and make the others like it, or we can arbitrarily bring everything to some convenient norm — say, 72. In

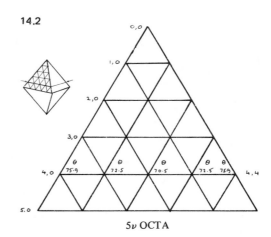

14.2

5v OCTA

any case, since the bottom row (5,0-5,5) will be discarded in the truncation process, and mirror symmetry takes care of the right half of the system once we have determined the left, this means, at most, recalculating just 6 values: 3,0/4,0; 4,0/4,1; 3,0/4,1; 4,1/4,2; 3,1/4,1; 3,1/4,2. Everything else is left as it was.

The icosa as usual requires a little more care. What Fuller calls the icosacap (the five triangles around the zenith) makes so shallow a saucer we are unlikely to want to use it by itself, let alone truncate it. Icosa truncations will cut through the 10 triangles that girdle the sphere's waist, the #2 triangles in the diagram we took into account with ellipsoids in Chapter 13.

Diagram 14.3 shows a *3v* Class I icosa opened out to display all twenty face triangles. A likely plane of truncation would follow line *A*, passing below the equator to give the so-called 5/8 truncation. Inspecting the *3v* Class I icosa tables, we see at once why amateur builders have gotten by with this truncation as it stands, using blocks to shore up the discrepant places: along the line 4,1-4,3, which is where we are cutting, the θ values luckily differ by only 2/3°. To achieve a perfect truncation, we have only to raise the θ of point 4,2 from 79.1877 to 80.1169 and refigure the associated chord factors. (We might elect instead to leave 4,2 alone and bring 4,1 and 4,3 into agreement with it, but then we should also have to alter 5,2 and 5,3 on the down-pointing adjacent triangles to make them match. It saves work to leave edge points alone when we can.)

So many *3v* icosa domes with uneven 5/8-sphere truncations have been built from *Domebook* data that full data for a smooth cutoff may be interesting. The basic chord factors are:

a	0,0/1,0	0.348615
b	1,0/1,1	0.403548
c	1,0/2,1	0.412411
d	1,0/2,0	0.403548

These are used for the five top (icosacap) triangles. They are also used for the #2 triangles that point downward below them; for a 5/8 truncation the down-pointing triangles are completed as far as 5,2/5,3. It is on the up-pointing #2 triangles flanking them that

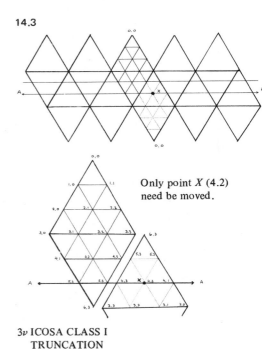

14.3

Only point *X* (4.2) need be moved.

3*v* ICOSA CLASS I
TRUNCATION

we move 4,2 to its new position *4,2,* making its θ equal to that of 4,1 and 4,3 (that is, 80.1168454). With this we obtain new chord factors for *4,2/4,1* and *4,2/5,2:*

| e | *4,2/4,1* | 0.412703 |
| f | *4,2/5,2* | 0.398734 |

These are used only for the members around the truncation perimeter; for the rest of the dome we use regular chord factors.

If we wanted a 3/8 truncation instead of a 5/8 truncation, study of the diagram shows that we should use a revised chord factor for *4,2/3,1* instead of *4,2/5,2.* It is

| g | *4,2/5,2* | 0.426214 |

EXAMPLE 2. We may remember that we left dangling in Chapter 13 the question of truncating a 2v icosa Class II ellipse for which we had calculated chord factors. One way of doing this is now obvious. We could change the θ of 3,0 to agree with that of 3,1 and calculate two new chord factors, 3,0/3,1 and 2,0/3,0. These would be 0.613085 and 0.378907, respectively. The latter figure provides for some rather stubby structural members, but the dome will sit flat. It will also curve in slightly near the bottom, since the truncation θ (100.8123170) lies below the 90° line.

Or we might move both 3,0 and 3,1 up to 90° (their coordinates would be 0,90 and 36,90, respectively) and refigure four distances: 2,0/3,0; 2,0/3,1; 3,0/3,1; 2,1/3,1. This would truncate the ellipse precisely at its zone of maximum radius, though all the base members would be somewhat short.

A better way to truncate Class II spheres introduces us to a new principle:

Truncation by Rotation

While we were struggling with the last example, the reader may have been wondering what happened to the great circles to which Class II breakdowns give access. The answer is that they are ready and waiting, but not while a vertex is at the zenith. Diagram 14.4 shows this; if we rotate the structure till an edge is on top we

14.4

ICOSA and FIFTEEN GREAT CIRCLES

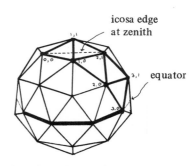

2v ICOSA: EDGE ZENITH

obtain equatorial great circles, twice interrupted in their circuit of the sphere.

Since the whole coordinate system hangs together, what we need is a way of rotating it as a whole until the midpoint of an icosa edge moves to the zenith.

The midpoint of a Class II icosa edge is 36,31.7174744. This is listed in the Chapter 21 tables as Point 2,1 of a 4ν Class II breakdown. If this isn't clear, consult Diagram 10.6, which shows the icosa edge passing down the middle of a Class II triangle. We need a way of telling what the other points become when this point becomes 0,0. We use the rotation formulae*, which tell us that when ϕ,θ moves to 0,0, then θ_1 will move to:

$$\theta_2 = \arccos\,\{\cos\theta_1 \cos\theta + \sin\theta_1 \sin\theta \cos(\phi-\phi_1)\},$$

[Eq. 14.1]

and ϕ_1 will move to:

$$\phi_2 = \arcsin\,\{\sin\theta_1 \sin(\phi-\phi_1)/\sin\theta_2\}. \qquad \text{[Eq. 14.2]}$$

Let us try this out with a 2ν Class II sphere. The Class II Method 1 tables give us the following coordinates for an icosa:

0,0	0	0
1,0	0	37.37736811
1,1	72	37.37736811
2,0	0	79.18768300
2,1	36	63.43494883
3,0	0	116.5650512
3,1	36	100.8123170

We can see from Diagram 14.5 that when a 2ν structure is rotated in the manner we contemplate, 3,0 will lie on the equator and 3,1 somewhere above it. Clearly these are the only points that will have anything to do with a 2ν truncation, so we ignore all the others and feed these two into the rotation formulae. We obtain, if we use 36 for ϕ and 31.7174744 for θ, the following new coordinates:

*Derived by Professor George Owen.

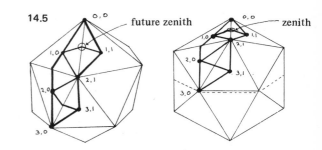

14.5

| 3,0 | -31.717474 | 90 |
| 3,1 | 0 | 69.0948426 |

We will have a truncation if we create a new vertex, 0,90, for the center of the diamond we are slicing and join both these points to it. The point can be called *x,* and the chord factors will be:

| 3,0/x | 0.546533 |
| 3,1/x | 0.362843 |

This will give us an edge-zenith sphere with a truncation utilizing one of its inherent great circles. The regular chord factors will take care of all the other structural members.

This particular problem could be solved in perhaps less time geometrically. But at higher frequencies than 2v, coordinate rotation is the way to solve Class II truncation problems. The diamond being sliced by the truncation plane has always the same boundaries; at higher frequencies we are simply inserting more new members (Diagram 14.6).

It would be quite possible to handle the Class II ellipse in the same way. We should first rotate *all* the coordinates, then obtain radii, chord factors, and truncation members. The method is not recommended for three reasons:

1. It is a great deal of work.

2. An ellipsoid with an edge zenith is an awkward looking structure.

3. By the time we have finished the coordinates have undergone so much manipulation that rounding-off errors make the fifth place of the chord factors wobbly.

Novel Truncation Planes

Coordinate rotation has other uses. Occasionally, examining a model, we discern a lesser-circle truncation plane made available by moving an unusual point to the zenith.

Diagram 14.7 shows a 3v Class I octa sphere. It has the usual octa equators, suitable for hemispheres. Rotating a model, however, till the central vertex of an octa face (2,1) is at the zenith, we

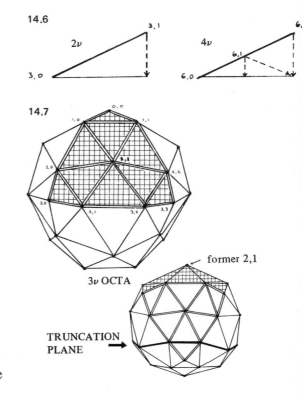

14.6

2v 4v

14.7

3v OCTA

former 2,1

TRUNCATION PLANE

notice two irregular lesser circles, resembling those of the 3ν icosa. Truncation along the lower one of these is an interesting possibility. The resulting structure is somewhat taller than a hemisphere and has three quadrangular skylights symmetrically disposed around the zenith.

Our first step, clearly, is to make the coordinate system correspond with the way we are viewing the model, by moving 2,1 to the zenith. We won't, of course, be calculating the result of moving every point — just the ones that will be affected when we straighten out that lesser circle. Diagram 14.8 shows the situation. Rotating the sphere toward the eye until the 2,1 point on the *far* side arrives at the zenith, we discover that the array 2,0-2,1-2,2 on the face nearest us has moved down to the desired truncation plane, which is irregular because 2,1 is misaligned with its neighbors. To straighten out the truncation plane we have only to equalize the θ values these three points assume in their rotated position. We shall be altering the θ occupied by a shifted 2,1, and shall need its revised distances from a shifted 2,0 and a shifted 1,0. So we need rotate only these three points.

The coordinates of the point on the far side of the sphere, which we propose to move to the zenith, are 225, arctan$\sqrt{2}$. (Its θ is the same as that of 2,1, and its ϕ is 180° greater.) Inserting these figures into Equation 14.1 and Equation 14.2 along with the coordinates of the three points to be recalculated, we obtain their new coordinates:

1,0	19.1066053	75.0367825
2,0	40.8933947	104.9632174
2,1	0	109.4712206

As we might have expected, the new θ of 2,1 is relatively close to that of 2,0 (and by symmetry, of 2,2). To effect the truncation, we equate them, taking the edge-point value (104.9632174) as the norm. Using this new value for *2,1* we obtain the following revised chord factors:

2,0/2,1	0.674981
1,0/2,1	0.607864

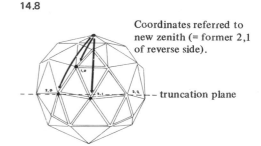

14.8

Coordinates referred to new zenith (= former 2,1 of reverse side).

truncation plane

Being very careful to notice what goes where (Diagram 14.9), the builder now employs these values in zones adjacent to the truncation plane and constructions a rotated/truncated 3ν octahedron with 2,1 at the zenith. The disposition of the square skylights is pretty.

For a more tightly woven mesh, we might double the frequency of this. For a 6ν we would straighten out the row 4,0-4,1-4,2-4,3-4,4. Mirror symmetry permits us to do this by attending only to 4,1 and 4,2 plus the points connected with them, 3,0 and 3,1. But (see Diagram 14.10) the 2,1 point on the adjacent, down-pointing octa face is now located on the truncation plane and must be aligned with its neighbors. When the octa was in its unrotated position, this was 5,1 on the next face clockwise; take the ϕ of 5,1, add 90° to get round to the next face, and relocate the result like the other points that get relocated. Then give it the same θ as the rest of the truncation plane and obtain new chord factors.

The square windows on the 6ν version will be smaller than on the 3ν. Should we prefer the larger size, we could omit 1,0/1,1, 1,0/2,1 and 1,1/2,1, then figure 0,0/2,0 and 2,0/2,2 as a single long chord.

Should we like this well enough to attempt an elliptical version, we should need to move more points: Diagram 14.11 shows where they are located.

Zenith Altitude and Floor Radius

Having truncated a dome we shall want to know how high its zenith point will be and how large its floor. Diagram 14.12 shows the derivation of two very simple formulas. The zenith altitude is simply:

$$\text{altitude} = r - \cos\theta, \qquad \text{[Eq. 14.3]}$$

and the floor radius is given by:

$$\text{floor radius} = r \sin\theta. \qquad \text{[Eq. 14.4]}$$

In these expressions θ is of course the θ of the truncation plane, and for spheres $r = 1$. So for the rotated 3ν octa discussed pre-

14.9

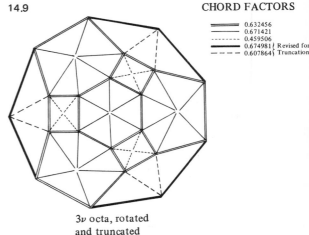

CHORD FACTORS

═══	0.632456
───	0.671421
-------	0.459506
▬▬▬	0.674981 } Revised for
─ ─ ─	0.607864 } Truncation

3ν octa, rotated
and truncated

14.10

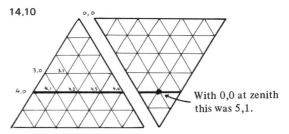

With 0,0 at zenith
this was 5,1.

14.11

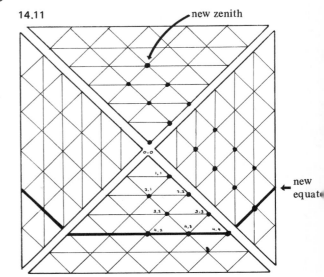

new zenith

new equator

viously, the zenith altitude is 1 – cos 104.9632174, or 1.258199, which means we have gained 25 percent more zenith headroom than a hemispherical truncation would have permitted, and the floor radius is sin 104.9632174, or 0.966092. The floor area is approximately πr^2, or 2.932153 ("approximately" because we are pretending a polygon is a circle); since the floor of a hemisphere (radius = 1) would be 3.141593, we have gained our extra headroom at a cost of 6.66 percent in floor area.

Should we be analyzing elliptical contours, we should need slightly more complicated expressions: if r_z is the radius of the zenith point, and r_t is the radius at the truncation plane, then

$$\text{altitude} = r_z - r_t \cos \theta. \qquad [\text{Eq. } 14.5]$$

$$\text{floor radius} = r_t \sin \theta. \qquad [\text{Eq. } 14.6]$$

If the floor is itself elliptical, there will be numerous values of r_t; we should probably be interested in the largest and the smallest, which would give us the length and width of the floor.

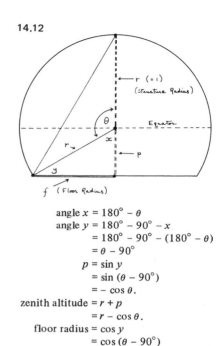

14.12

$$\text{angle } x = 180° - \theta$$
$$\text{angle } y = 180° - 90° - x$$
$$= 180° - 90° - (180° - \theta)$$
$$= \theta - 90°$$
$$p = \sin y$$
$$= \sin (\theta - 90°)$$
$$= - \cos \theta.$$
$$\text{zenith altitude} = r + p$$
$$= r - \cos \theta.$$
$$\text{floor radius} = \cos y$$
$$= \cos (\theta - 90°)$$
$$= \sin \theta.$$

15. An Advanced Problem

Among the means at our disposal we may now list the following:

- Writing the ϕ,θ coordinates for Class I or Class II breakdowns of icosa, octa, or tetra polyhedra (Chapter 12, data in Part 3).
- Obtaining the distance (chord factor) between any two points on a spherical contour (Equation 9.2).
- Writing the third coordinate, r, for any θ when the floor plan is circular and the silhouette is elliptical or superelliptical (Equations 13.1, 13.3, 13.4).
- Writing the third coordinate, r, for any ϕ, θ when both floor plan and silhouette are elliptical or superelliptical (Equation 13.7).
- Correcting for uneven expansion (Equations 13.2, 13.8, 13.9).
- Obtaining the distance (chord factor) between any two points on a nonspherical contour, once we have their three coordinates ϕ, θ, and r (Equation 9.1).
- Inserting truncation planes at will (Chapter 14).
- Rotating the coordinate system to bring any point to the zenith (Equations 14.1, 14.2).

We may now want to tackle a sophisticated problem that tests our resources. The harvest of insights seems to be worth the few pages we shall need to work it out in full. The reader who wants to test his command of the equations can spot-check any result at any point.

Near the end of Chapter 13 we were considering a superelliptical floor plan and contour, with a supercircular cross section. This offers the advantage of considerable floor space, with headroom maintained well out from the center. Because it is symmetrical by quadrants, it invites the use of the octahedron, one face triangle of which yields one quadrant exactly.

Unfortunately, it compounds a defect of the octahedron, deficient spherical excess at the principal vertices. If we are standing inside an octahedral geodesic hemisphere, a four-way vertex is directly overhead. If the breakdown is 4-frequency, the radials of this vertex will have chord factors of 0.141778, bounded by a square whose sides are 0.200000. These are almost right-angled triangles; the four angles meeting at the vertex are each 89.7121°, alarmingly close to the 90° which would make this vertex perfectly flat. Octahedral spheres are feasible because the small flat area is fully surrounded by vertices whose departure from flatness is correspondingly more emphatic. There is nothing wrong with flat vertices properly distributed. Fuller himself pioneered the *hex-pent* configuration, which replaces pentagonal and hexagonal arrays with flat plates, in effect flattening numerous vertices throughout the system and relying on surrounding tensional grids to hold things together, and numerous hex-pent radomes have been withstanding Arctic winds on the DEW line for many years.

But once we introduce the superspherical or superelliptical contour, we find that quasiflatness is creeping a considerable distance out from the zenith. Not only is the zenith of the octa nearly flat, but so are the four vertices around it. Try the icosahedron. The angles around the pent vertex of a $4v$ sphere are each 71.3316°, so five total 356.658°. The four octa angles totaled 358.848°, so the icosa is helpfully less flat.

Now try it with the superellipse. We shall be using the same parameters throughout this chapter: floor expansion, 1.5; roofline expansion, 1,0; both exponents, 2.5. Above the superelliptical floor, we have a supercircle profile in one cross section, a superellipse profile in the other. So if the pent vertex overhead has a radius of 1, the 5 vertices surrounding it have radii greater than 1, pushing the ends of the radials out toward flatness. Calculating radii and chord factors, we find that the 5 pent angles are each 71.985°, or 359.925° for the set — virtually flat.

Moreover, we have quadrant symmetry, and the icosa with its 72° radials will be awkward to quarter.

The solution, discovered by Bill Wild of Vail, Colorado, whose results appear without commentary in *Shelter* (Bolinas, Ca.: Shelter Publications, 1973) p. 126, is to move the icosa vertices away from the zenith. Move an edge to the zenith, and let it follow the supercircular profile, so that the pent vertices at its ends will be carried down to where curvature is accentuated. Then instead of having less spherical excess than when the envelope is a sphere, they will have more. And overhead there will be no vertex at all to suffer that undue flatness (Diagram 15.1).

There is a bonus. With an edge up, the icosa will quarter neatly, the cuts running along that edge and straight down the centers of face triangles.

To achieve this we do the following:

1. Discover how much of the icosahedron we shall need to account for, to include all points in a quadrant of a hemisphere.
2. Write the coordinates of these points.
3. Rotate the system of coordinates till the center of an icosa edge is at 0,0.
4. Obtain the radii of all points in this new position.
5. Obtain the chord factors.
6. Attend to any truncation problems.

Quartering the Icosa

Diagram 15.2 shows the situation. The icosahedron is edge-upward, and the fifteen great circles are marked on it. Their 90° intersections at face-triangle edges give us just the demarcations we want; as the diagram shows, a full icosa triangle plus halves of three others will exactly cover one quadrant of a hemisphere. (This makes 2 1/2 triangles, and since 4 × 2 1/2 = 10, our result agrees with the fact that 20/2 = 10 icosa triangles would constitute a hemisphere.)

Writing the Coordinates

Look at Diagram 15.3. The line 0,0-4,0 is the edge that will move to zenith. The full triangle we want is the *second* icosa triangle in the tables, 4,0-4,4-8,4. Of the upper icosa triangle (0,0-4,0-4,4) we

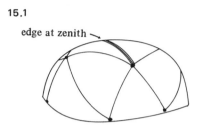

15.1

edge at zenith

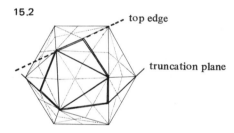

15.2

top edge

truncation plane

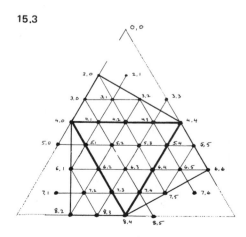

15.3

need only half, as shown. Since two struts pass across its upper boundary, we also need points 2,1 and 3,3. All these points are obtained directly from the tables.

To obtain coordinates for the part-triangles 4,0-8,2-8,4 and 4,4-8,4-6,6 we remember how the icosa is put together (see Diagram 15.4). These are also second triangles, but up-pointing. Any point in them will be the same distance from the bottom icosa vertex ($\theta = 180°$) as the corresponding point in the usual second triangle is from the top vertex ($\theta = 0°$). So we identify corresponding points and subtract from 180°. Thus 7,2 corresponds to 5,2; subtracting the θ of 5,2 from 180°, we obtain a θ for 7,2: 106.0450570°.

To obtain the ϕ values of these points, we note that symmetry places them left and right of icosa-triangle centerlines exactly as the corresponding points in the second triangle are placed left and right of its centerline. Its centerline is at $\phi = 36°$. The up-pointing triangle centerlines are at $\phi = 0°$ and $\phi = 72°$. Since 5,2 is 36° – 26.2676986° = 9.7323014° from the 36° centerline, the ϕ of 7,2 is 0 + 9.732014 = 9.7323014°, and the ϕ of 7,5 is 72 – 9.7323014 = 62.2676986°.

Again notice that edge-crossing struts will require us to write some coordinates outside these boundaries. Since 5,0 and 7,1 are to the left of the $\phi = 0$ line, their ϕ values are negative.

Table 15.1 lists all of the coordinates we require.

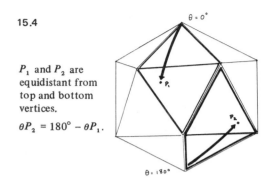

15.4

P_1 and P_2 are equidistant from top and bottom vertices.

$\theta P_2 = 180° – \theta P_1$.

TABLE 15.1   *Coordinates from Table*

	ϕ	θ		ϕ	θ
2,0	0	31.7174744	5,1	8.7723551	75.4545635
2,1	36	26.5650512	5,2	26.2676986	73.9549430
3,0	0	48.8895123	5,3	45.7323015	73.9549430
3,1	22.3861776	43.6469271	5,4	63.2276449	75.4545634
3,2	49.6138225	43.6469271	5,5	80.7723551	75.4545634
3,3	72	48.8895123	6,1	0	90
4,0	0	63.4349488	6,2	18	90
4,1	16.0353713	59.8480232	6,3	36	90
4,2	36	58.2825256	6,4	54	90
4,3	55.9646288	59.8480232	6,5	72	90
4,4	72	63.4349488	6,6	90	90
5,0	–8.7723551	75.4545635	7,1	–9.7323014	106.045057

TABLE 15.1 *Coordinates from Table (continued)*

	ϕ	θ			ϕ	θ
7,2	9.7323014	106.045057		8,2	0	121.717474
7,3	27.2276449	104.545437		8,3	19.9646287	120.151977
7,4	44.7723551	104.545437		8,4	36	116.565051
7,5	62.2676986	106.045057		8,5	52.0353713	120.151977
7,6	81.7323014	106.045057				

Rotating the Coordinate System

We are now going to move the midpoint of the edge 0,0-4,0. This midpoint is 2,0, and since it will now occupy the zenith, its ϕ and θ will be 0° and 0°. So we want to know what the ϕ and θ of each point will be when the present 0, 31.71747438 has been moved to 0,0. Equation 18.1 and Equation 18.2 give us the answers, with tedium but not difficulty. One caution is needed: we must watch that we assign ϕ to the right quadrant. Diagram 15.5 shows the new boundaries: normally all values will lie between 0 and 90. Thus 8,5 and 7,6 are the only points below the equator. Here Equation 14.1 will give us the proper θ values. Point 2,1, however, lies beyond $\phi = 90°$, and here we confront the fact that Equation 14.2 yields its answer as an arcsine, in this case arcsine 0.8506508. The unthinking answer is 58.2825260. But we need something greater than 90°; the value we want must be in the second quadrant, 121.7174740. Similarly, the ϕ values for 5,0 and 7,1 must be negative but not a great deal less than zero: they are in fact the negative values of their counterparts 5,1 and 7,2. An eye kept on the diagram's physical realities will prevent mistakes.

Table 15.2 shows the rotated coordinates. To prevent confusion the points keep their old labels.

15.5

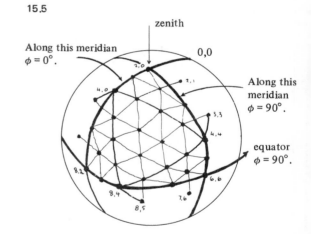

TABLE 15.2 *Coordinates after Rotation*

	ϕ	θ	r
2,0	0	0	1.0
2,1	121.717474	18	1.0404647
3,0	0	17.1720379	1.0252556
3,1	58.2825256	18	1.0404647

TABLE 15.2 *Coordinates after Rotation (continued)*

	ϕ	θ	r
3,2	90	31.7174744	1.1279268
3,3	101.640723	47.0219324	1.2703153
4,0	0	31.7174744	1.0583602
4,1	28.3399674	30.2099053	1.0748489
4,2	58.2825256	36	1.1494618
4,3	78.3592768	47.0219324	1.2703153
4,4	90	58.2825256	1.3854747
5,0	-12.2183198	44.2282383	1.0818729
5,1	12.2183198	44.2282383	1.0818729
5,2	35.8897547	46.5129225	1.1529653
5,3	58.2825255	54	1.2935799
5,4	75.5495629	63.7133234	1.4280485
5,5	90	72.8279622	1.4865500
6,1	0	58.2825256	1.0583602
6,2	20.9051575	60	1.1018176
6,3	40.5007050	64.8287379	1.2210322
6,4	58.2825256	72	1.3843498
6,5	74.5495630	80.6502964	1.4946122
6,6	90	90	1.5
7,1	-9.6937239	74.7598376	1.0332939
7,2	9.6937239	74.7598376	1.0332939
7,3	27.1333780	76.1809908	1.1104788
7,4	43.5723494	81.5109204	1.2415088
7,5	58.2825256	90	1.3854747
7,6	74.5495629	99.3497034	1.4946122
8,2	0	90	1.0
8,3	17.1720378	90	1.0386996
8,4	31.7174744	90	1.1279268
8,5	43.5723495	98.4890796	1.2415088

If we intend to alter the coordinate symmetry with the help of Equations 13.8 and 13.9, now is the time to do it. To shorten the example we'll skip this option, remarking only that since the floor is being expanded along the $\phi = 90°$ axis, both equations would require the modification explained previously.

Tabulating the Radii

Again the work is tedious but perfectly straightforward. There is only one thing to remember: the direction of the long axis of the floor. With a vertex at the zenith — the usual situation — this is unimportant, but here we want to be sure that the vertexial edge runs along the supercircular rooftree, not the superelliptical one. This is chiefly to work the pent centers at the ends of this icosa edge as far down into the zone of curvature as we can. (It also happens to yield a tighter clustering of chord factors.) Now Equation 13.5 as it stands will expand the floor along the $\phi = 0$ axis, which is 90° from where we want it. The easiest way to swing things 90° is to alter Equation 13.7 very slightly; we have only to interchange sin ϕ and cos ϕ, and Equation 13.7 will swing the floor plan as we want, and generate the radii as tabulated in the right-hand column of Table 15.2.

Chord Factors

This is the ungrateful case where symmetry fails us on any scale smaller than 1/8 sphere; hence, we need chord factors for all the struts in the dome quadrant. A good way not to miss any is to mark each strut red on a diagram as its value is computed.

Equation 9.1 will give the values in Table 15.3. They do not correspond to those shown by Mr. Wild in *Shelter* because his expansion was only 1.25, while ours is 1.5. A model will tell us whether we like the looks of it; if not, we can go back to Mr. Wild's stubbier form. (Refigure radii, refigure chord factors.)

TABLE 15.3   *Chord Factors*

2,0/3,0	0.303388	4,1/5,1	0.318453	5,2/6,2	0.356325
2,0/3,1	0.321691	4,1/5,2	0.337055	5,2/6,3	0.391483
2,1/3,1	0.338067	4,2/4,3	0.380510	5,3/5,4	0.422293
3,0/3,1	0.304581	4,2/5,2	0.360196	5,3/6,3	0.414748
3,0/4,0	0.265806	4,2/5,3	0.407824	5,3/6,4	0.428406
3,0/4,1	0.313876	4,3/4,4	0.355079	5,4/5,5	0.434129
3,1/3,2	0.362693	4,3/5,3	0.378274	5,4/6,4	0.422092
3,1/4,1	0.313364	4,3/5,4	0.427795	5,4/6,5	0.435413
3,1/4,2	0.359097	4,4/5,4	0.358665	5,5/6,5	0.439095
3,2/4,2	0.356896	4,4/5,5	0.377147	5,5/6,6	0.446072
3,2/4,3	0.380223	5,0/5,1	0.319414	6,1/6,2	0.340641
3,3/4,3	0.375050	5,1/5,2	0.336542	6,1/7,2	0.340709
4,0/4,1	0.270543	5,1/6,1	0.316039	6,2/6,3	0.381975
4,0/5,1	0.271939	5,1/6,2	0.326608	6,2/7,2	0.340793
4,1/4,2	0.340173	5,2/5,3	0.420986	6,2/7,3	0.330394

TABLE 15.3   *Chord Factors (continued)*

6,3/6,4	0.438331	7,2/7,3	0.324803	8,3/8,4	0.288206
6,3/7,3	0.360343	7,2/8,2	0.319773		
6,3/7,4	0.363212	7,2/8,3	0.305180		
6,4/6,5	0.463349	7,3/7,4	0.370600	TRUNCATION	
6,4/7,4	0.416692	7,3/8,3	0.325110	6,5/7,6	0.243588
6,4/7,5	0.433297	7,3/8,4	0.283893	6,6/7,6	0.412714
6,5/6,6	0.468491	7,4/7,5	0.412238	7,4/8,5	0.187107
6,5/7,5	0.480131	7,4/8,4	0.320415	7,5/7,6	0.410359
6,5/7,6	0.485629	7,4/8,5	0.366545	7,5/8,5	0.333335
7,1/7,2	0.335738	8,2/8,3	0.306762	8,4/8,5	0.300297

Truncation

Just a small problem. Though we are using an even-frequency icosahedron, the rotation to edge zenith has a great-circle equator that cuts through triangles. This is the same great circle that supplies broken equators for Class II breakdowns, and it is dealt with in the same way: we figure special members to close the gaps where they occur.

It turns out that we don't want 7,4/8,5 and 6,5/7,6 after all, unless to construct a sphere model. We want midpoints, *8,5* and *7,6* (italicized to denote a change) between points 8,4-7,5 and 7,5-6,6, and we want a total of 6 new members to join things up. The new coordinates are:

8,5	45.0000000	90	1.2494849
7,6	74.1412628	90	1.4910254

The ϕ values simply split the difference between the ones on either side of them. The θ values are manifestly 90°. The radii are obtained as usual. The chord factors for the truncation members are then:

7,4/8,5	0.187102	*6,5/7,6*	0.243588
8,4/8,5	0.300297	*7,5/7,6*	0.410359
7,5/8,5	0.333335	*6,6/7,6*	0.412714

The complete dome will require four members of each of these lengths.

This completes the work. Which angles, if any, we may need to calculate will depend on the envisaged method of construction. How to do this, and how much of it to do, are outlined in Chapter 17.

16. Space Frames

Very large domes keep the lengths of their members within bounds by using very large frequencies; as we shall soon see, a 32ν has been constructed. But very large frequencies engender chord factors that are very much alike; two typical values for a 32ν are 0.033004 and 0.031676; for a 100-foot radius the members corresponding to these differ in length by less than 1 1/2 inches (about 4 percent). But zero difference in length is the formula for flatness, so any vertex in a dome of this frequency is getting nearly flat. The tension system is losing mechanical advantage, and a little stretch here and there (elasticity of materials; inaccuracy of construction; play of joints) would permit a vertex to pop inward under load.

At how low a frequency this problem becomes important depends on materials and tolerances, but it is probably wise to be wary of anything above 8ν or 10ν.

Being wary means imparting a third dimension to the frame. The frames we have been discussing are (mathematically) of zero thickness: though wood or metal has a third dimension, the diagrams we have had wood and metal follow consist of widthless lines running from point to point. Thickness entails a second line closer to system center than the first. There are several ways of doing this elegantly and without redundant weight. All of them are reducible, essentially, to a *pair* of frames, outer and inner, trussed together.

Diamond Patterns

One especially elegant method, in use since at least 1960, is to make the skin integral with the trussing system. This is the principle of Fuller's patent on Laminar Geodesic Domes. It is based on the simple fact that the same set of points can be joined in two different ways to combine, in effect, two different breakdown systems. Diagram 16.1 shows the principle: instead of joining points A, B, C, D to make two triangles, ABC and BCD, we omit the connection BC and instead join A to D. The A-D connections follow a 2ν Class II breakdown, the others a 3ν Class I with a few members omitted. It could all be done with struts, though hub details would get complex; Fuller's solution was to insert a creased diamond as shown.

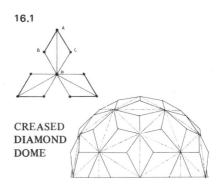

16.1

CREASED
DIAMOND
DOME

By building the entire dome in this way we can eliminate the strutted frame entirely and allow the joined edges of diamonds and the creases down the centers of the diamonds to take its place. Such a dome has in effect a single circumsphere, passing through all vertices, but two different inspheres, the 3ν and the 2ν; the difference between the radii of the latter is the virtual thickness we have conferred.

The diamond geometry is independent of dual radii. It was first worked out in the early 1950s, for a structure erected in Woods Hole, Massachusetts, and is interesting because it permits a 3ν structure with two diamond diagonals but only one edge length.*

Diagram 16.2 shows how. We want to make $0{,}0/1{,}0$ equal to $1{,}0/2{,}1$. We commence by noting that if $2{,}1$ (the centroid of the icosa face) has coordinates ϕ, θ while $1{,}0$ has coordinates ϕ_1, θ_1, then the distance $0{,}0/1{,}0$ will be (by Equation 9.2):

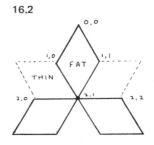

16.2

$$\sqrt{2 - 2 \cos \theta_1},$$

and the distance $1{,}0/2{,}1$ will be:

$$\sqrt{2 - 2 \left(\cos \theta \cos \theta_1 + \cos \phi \sin \theta_1 \sin \theta \right)}.$$

From the 3ν Class I table (Chapter 20) we learn that the coordinates of $2{,}1$, the centroid of the icosa face triangle, are 36, 37.3773681. We insert these values for ϕ and θ, equate the two expressions with one another, and solve this equation for θ_1, which proves to be 22.6904803.

*See Edward Popko, *Geodesics* (Detroit: University of Detroit Press, 1968), Figs. 15-19; but the geometric proof in his Figure 19 is full of misprints.

From this we derive a 3ν icosa breakdown using only 3 chord factors:

0,0/1,0		
1,0/2,1	0.393411	(all diamond edges)
1,0/1,1	0.453480	(fat-diamond diagonal)
1,0/2,0	0.313800	(thin-diamond diagonal)

These chord factors are for a dome made in the usual way, out of struts. To make it of creased diamonds, we need two more distances, the long diagonal common to every fat diamond (0,0/2,1) and the long diagonal of every thin diamond (2,1/4,2). These are 0.640852 and 0.713644, respectively.

A little experimenting with cardboard diamonds will show that once the edges are correctly cut, the angles of crease are self-adjusting. For stiffer material, for instance sheet metal creased in a brake, we would want to know the crease angles in advance. As Diagram 16.3 shows, they are easily calculated: 166.8° for the fat diamond, 142.52° for the thin.

If we make a cardboard model, or even fit just a few cardboard diamonds together, we can easily see that, whereas the material is thin, the geometry of the shell confers *virtual* thickness, interesting but unnecessary for a frequency this low. The easiest way to extend the principle to higher frequencies where it can be of some use is to work from a Class II breakdown. Diagram 16.4 shows the derivation of the diamonds for a 4ν and Diagram 16.5 shows how they fit together. Again, a second breakdown has appeared: the creases outline a 2ν Class I. Coordinates need not be tampered with, but simply taken from the Class II table (Chapter 21 or Chapter 23). Again two diamonds suffice.

It should now be clear how higher frequencies would be dealt with. Make a Class II breakdown diagram; mark symmetries (that is, equal chord factors: colored pencils are useful); outline the diamonds; calculate the necessary chord factors, remembering to include the long diagonal of each diamond. If the crease angle is needed, obtain it as shown in Diagram 16.3. The creases will follow a Class I breakdown at half the chosen frequency.

Both of the Class II breakdown methods discussed in this book

16.3

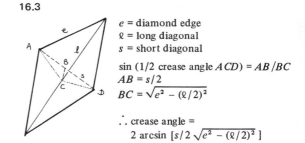

e = diamond edge
ℓ = long diagonal
s = short diagonal

$\sin(1/2 \text{ crease angle } ACD) = AB/BC$
$AB = s/2$
$BC = \sqrt{e^2 - (\ell/2)^2}$

∴ crease angle =
$2 \arcsin[s/2 \sqrt{e^2 - (\ell/2)^2}]$

16.4

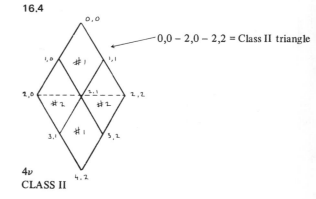

0,0 – 2,0 – 2,2 = Class II triangle

4ν
CLASS II

yield two diamonds at 4*v*. At higher frequencies Method 3 displays its usual economies, the number of different diamonds being just half the frequency count (four for 8*v*, eight for 16*v*). Method 1 requires six diamonds for 8*v* and twenty for 16*v*, the number of different components, as usual, increasing rapidly.

Trussed Frames

The other principal method of lending stiffness to a shell of high frequency (low local curvature) is to use an outer frame and an inner frame and truss them together.

The American Society for Metals (ASM) dome in Cleveland (Popko, figures 69-72) consists of two shells of identical geometry, each 24*v* Class II, the radius of the inner one 30 inches less than the radius of the outer, joined together at the vertex points by 30-inch radials. How something like this could be designed is obvious: two sets of strut lengths are figured, using the same chord factors but two slightly different radii.

To save weight (and lend grace) numerous struts are omitted, leaving a pattern of hexagons which are then stabilized tensionally, the tension cables replacing the omitted struts (Popko, figure 71, shows connector details).

Or the two shells need not be identical. The Montreal Expo '67 dome, a *tour de force* of design by Fuller and Sadao Inc., in effect trusses a 16*v* Class I outer grid to a 32*v* Class II inner grid. As we learned from our discussion of diamonds, both classes can be derived from Class II coordinates by choosing different points to interconnect. Diagram 16.6 shows the principle, using a lower pair of frequencies (4*v* and 8*v*). The hexagon pattern is the inner shell, obtained by joining selected 8*v* Class II vertices. The triangle pattern is the outer shell, obtained by joining vertex pairs as shown. Finally the missing struts in the hex-pattern inner shell are replaced by truss members which rise from the hex corners to meet, six at a time, at the outer shell vertices. The result is an exceedingly rigid space frame, virtually a spherical octet truss.

As an exercise, we may quickly obtain every member length in our 4*v*/8*v* example. Let us assume a ratio of 100/95 for the radii of the two shells and select Class II Method 3 for its characteristic

16.5

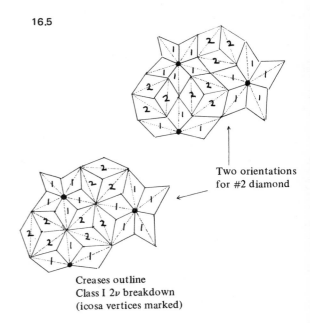

Two orientations for #2 diamond

Creases outline
Class I 2*v* breakdown
(icosa vertices marked)

16.6

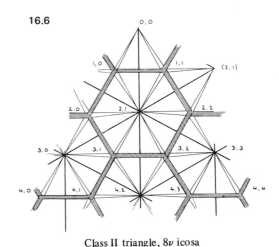

Class II triangle, 8*v* icosa

economies. We first see which 8ν chord factors we need; Diagram 16.7 shows that the hex and pent boundaries use only 3 lengths, *b, c,* and *h.* We extract these directly from the table in Chapter 23, and because this shell will have a radius of 0.95 we take 95 percent of the result:

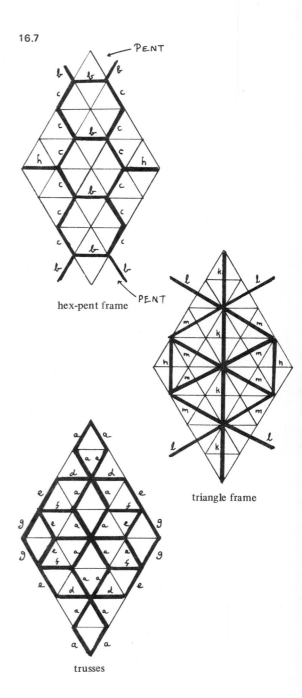

16.7

PENT

hex-pent frame   PENT

triangle frame

trusses

8ν Class II Method 3
(r = 0.95)

1,0/1,1	*b*	0.189484
1,0/2,0	*c*	0.158653
4,0/4,1	*h*	0.160239

We next obtain chord factors for the outer shell's triangulation, using Equation 9.2 and the coordinates from the table. This shell has a radius of 1, so the chord factors are written down exactly as they come from the calculator. (Just one detail: the strut from 2,1 which crosses the Class II triangle's edge goes to *2,1* in the next Class II triangle. This point has the same θ as 2,1 in the table, but its ϕ will be 72° higher — 108 instead of 36.) Only four values occur, and if our understanding of Class II Method 3 symmetry was insufficient to tell us this, we should soon find it out by experiment.

4ν Class I   (r = 1.0)

0,0/2,1	*k*	0.275904
2,1/2,1	*l*	0.321244
2,1/3,0	*m*	0.314751
3,0/5,1	*n*	0.264769

2,1/4,2 is the same as 0,0/2,1 and 3,0/4,2 is the same as 2,1/3,0; the other symmetries are obvious.

The truss members, finally. We know the spherical coordinates of every pair of points in which we are interested, and the standard formulae will give us the distance between them, regardless of the fact that they are not located on a common sphere. One point of each pair is located on an inner shell, one point on an outer. We simply use Equation 9.1 and both radii, 1.0 and 0.95. Of the eight member lengths in the 8ν Method 3 breakdown, hex and pent edges have already accounted for three; the trusses correspond to the other five.

truss members ($r_1 = 1$, $r_2 = 0.95$)

0,0/1,0	*a*	0.173343
2,0/2,1	*d*	0.197168
2,0/3,0	*e*	0.164721
3,0/3,1	*f*	0.187039
3,0/4,0	*g*	0.157577

These are used exactly like chord factors: we multiply by the radius of the outer shell to get the actual lengths of truss members. The lengths of inside shell members are also obtained by using the outside shell radius, the × 0.95 correction having been already built into the chord factors.

This very intricate two-shell trussed domical space frame uses members of just twelve different lengths. It's not a backyard project. The hubs would need careful study. Each hub on the outer shell receives twelve members, each hub on the inner shell six. Hub design would need angles which Chapter 17 tells how to obtain.

17. About Angles

A geodesic structure contains three main kinds of angular relationships about which we may need information: (1) *Axial.* This is the angle between a strut and the sightline from its end to system center. It is a function of *strut length* and *local radius.* If the system is spherical (all radii = 1) each strut length has its unique axial angle, the same at either end. If the system is nonspherical, varying radii may produce different axial angles at the two ends of a strut, and still different ones for the same strut length at different positions in the system. (2) *Face.* The corner angles of a face triangle. (3) *Dihedral.* The angles between adjacent triangular planes.

What angles we may want to know will depend on the method of construction we envisage. The builder of a wooden frame needs axial angles to know how to cut his struts where their ends meet. Hub design requires axial and face angles. To assemble pieces of paneling into a skin with snugly butted edges, we need dihedral angles. (The crease angle we encountered in our discussion of diamond panels was a species of dihedral, easy to calculate because the diamond was highly symmetrical.) Detailed information follows.

Axial Angles

Diagram 17.1 shows that when the system is spherical the axial angles are at the base of an isosceles triangle whose apex is the central angle subtended by a strut. So each axial angle is 1/2 (180° – central angle). And we can derive the central angle from the chord factor, which by definition is twice the sine of half the central angle. Combining these observations, we find that for a spherical system,

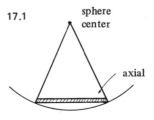

17.1   sphere center

axial

$$\text{axial angle} = \text{arc cos (chord factor}/2)$$

$$[\text{Eq. } 17.1]$$

If the system is nonspherical, we need, in addition to the strut's chord factor, the radii at its two ends, which will be in our work sheets because obtained them when we were calculating the chord factor in the first place. Diagram 17.2 shows the triangle we are now solving. If the strut is BC, its chord factor a, and the radii at ends B and C are R_b and R_c, respectively, then,

17.2

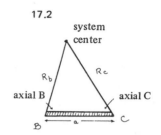

$$\text{axial } B = \text{arc cos} \left[(a^2 + R_b^2 - R_c^2)/2aR_b \right].$$

$$[\text{Eq. } 17.2]$$

After which, more simply,

$$\text{axial } C = \text{arc sin} \left[(R_b \sin \text{Axial } B)/R_c \right].$$

$$[\text{Eq. } 17.3]$$

Face Angles

From the freshman trig text: given the three sides of a triangle, to find its three angles. The sides are simply the chord factors of the struts that bound a face. In Diagram 17.3, if the triangle ABC has sides a, b, and c, then

17.3

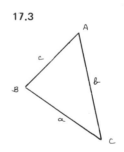

$$\text{face angle } A = \text{arc cos } (b^2 + c^2 - a^2)/2\,bc. \quad [\text{Eq. } 17.4]$$

We can use this result to find the next angle more quickly:

$$\text{face angle } B = \text{arc sin } (b \sin \text{face angle } A)/a.$$

$$[\text{Eq. } 17.5]$$

And the last one more quickly still:

$$\text{face angle } C = 180° - \text{face angle } A - \text{face angle } B.$$

$$[\text{Eq. } 17.6]$$

Dihedral Angles

The existence of an elegant procedure seems plausible, but so long as it continues to be elusive, we can rely on trigonometric brute force. Diagram 17.4 shows what we are trying to accomplish. We want the dihedral angle between triangle *ABC* and triangle *DBC*. A "normal plane" *OBC* passes through their common edge and the system center, and it is not too difficult to obtain the angle at which each triangle meets this plane. We can then obtain the dihedral by adding these two angles together.

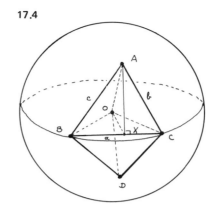

Start with triangle *ABC.* We know its three sides (chord factors) and the radii of its three vertices (= 1 for spheres). We know the face angle at *B* (or know how to obtain it — see above). We also know or can obtain the axial angle at *B.* A perpendicular *AX* to the side *BC* marks off segment *BX*, the length of which will be $c \cos B_f$ (B_f being the face angle). In the triangle *XBO* we now know 2 sides, *BX* and the radius *BO*; the angle included between them will be the axial angle B_a, which we also know how to obtain. So we can find side *OX*, which is also a side of triangle *AXO.* The side *AX* of this triangle is $c \sin B_f$, and its third side is the radius at *A*. So we can calculate what we want to find, the angle *AXO.* Using data for triangle *DBC,* we repeat the process to obtain angle *DXO,* and add these results. The sum is the dihedral.

When all radii = 1 (the spherical case) this procedure shakes down to:

$$\text{angle } AXO = \arccos \left(c - a \cos B_f / 2 \sin B_f \sqrt{c^2 \cos^2 B_f - ac \cos B_f + 1} \right).$$

[Eq. 17.7]

This method is messy but exact. To see how often we must go through it, we examine each of the different strut lengths in a breakdown diagram to see how many different pairs of triangles meet across it. Each strut is the ridge of a dihedral angle, and if our system is spherical the dihedrals along identical boundaries between like pairs of triangles will be identical.

Diagram 17.5 shows all the triangle types in the 3ν rotated and truncated octahedron we designed in Chapter 14. Its forty-five triangles are of just four types, the ones that group in hexagons, the ones that group in squares, and the two special ones produced by the truncation process. Of the five different strut lengths, *a* occurs on four kinds of boundaries: between △1 and △1; between △1 and △2; between △2 and △3; between △1 and △4. Length *b*

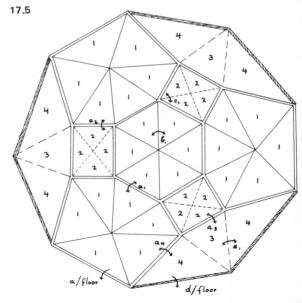

Despite distortions in the drawing, all triangles with same boundary sequence have the same shape.

An example of each dihedral type is marked.

occurs only between $\triangle 1$ and $\triangle 1$; length c only between $\triangle 2$ and $\triangle 2$. Length d never occurs between pairs of triangles, only between a triangle and the floor (more about this later). Length e is found only between $\triangle 3$ and $\triangle 4$.

So this structure contains just seven dihedral angles, not an excessive number to figure if our method of construction should require them. The four triangles will have a total of twelve face angles, and the five strut lengths will have five axial angles. The complete set is tabulated herewith (Table 17.1) for the reader to check as an exercise.

Table 17.1   *3ν Octa, Rotated and Truncated*
              (complete angular data)

	Chord Factor	Axial			
a	0.632456	71.565037		Face de	58.799355
b	0.671421	70.384251		Face ae	65.905155
c	0.459506	76.717470			
d	0.674981	70.275945		Dihedral a 1 1	154.208383
e	0.607864	72.306078		Dihedral a 1 2	166.094599
				Dihedral a 2 3	169.081500
$\triangle 1$:	Face ab	61.901864		Dihedral a 1 4	155.487605
	Face bb	56.196272		Dihedral b 1 1	158.079453
$\triangle 2$:	Face ac	46.512901		Dihedral c 2 2	154.688973
	Face cc	86.974198		Dihedral e 3 4	156.048901
$\triangle 3$:	Face ae	58.652439			
	Face ee	62.695121		Wall / floor, a	92.067409
$\triangle 4$:	Face ad	55.295491		Wall / floor, d	95.739409

Note: "Face ab" means the face angle between struts a and b. "Dihedral a 1 1" means the dihedral angle between two #1 triangles where they meet along side a. For triangle numbering see Diagram 17.5.

A word on the meeting of panel with floor (along strut d and sometimes strut a in this example). This constitutes a special kind of dihedral angle. Should we want to know it, we divide the problem as before into two parts. We use Equation 17.7 to find the angle between the triangular panel and a "normal plane" through system center. If we are dealing with a hemisphere, the floor occupies this plane and the calculation is finished. If we are not, we need also the angle between the normal plane and the floor.

Diagram 17.6 shows that this follows directly from the θ of the truncation plane: it is simply $\theta - 90°$. We add this to what Equation 17.7 gave us; the result will be the angle between panel and floor, as viewed (like all dihedral angles) from *inside* the structure — an acute angle if the structure is less than a hemisphere, obtuse if it is more. (The acute-obtuse correction is supplied automatically by the fact that $\theta - 90°$ will be negative if the structure is less than a hemisphere, so the net effect of adding it will be to diminish the dihedral.) In our rotated/truncated octa example, the truncation plane θ is 104.963218°, so we add 14.963218° to the angle between bottom triangles and the normal plane, giving us dihedral angles of slightly more than 90°. At 14° below the equator, a 3ν octa structure is just commencing to display outward curvature.

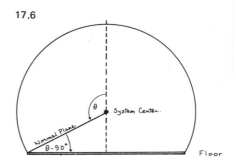

17.6

Nonspherical dihedrals remain to be considered. It is a comfort to know that they are rarely needed. They are tedious to arrive at, both because varying radii complicate the expressions, and because there are apt to be so many different ones. Anyone engaged in the design of a high-frequency nonspherical geodesic structure presumably has a budget sufficient to rent computer time if dihedral angles are needed. What will need to be gotten through the computer's head is a procedure essentially like that which yielded Equation 17.7, complicated however by the fact that the three radii of a triangle's vertices are no longer equal. The discussion that accompanied Diagram 17.4 may be reread with this fact in mind.

By staying calm we arrive at the following procedure:

1. Obtain the axial angle B_a with Equation 17.2
2. Obtain length OX as follows:

$$OX = \sqrt{c^2 \cos^2 B + R_b{}^2 - 2c \cos BR_b \cos B_a}.$$

[Eq. 17.8]

3. Obtain Angle AXO as follows:

angle $AXO = \text{arc cos}\ (c^2 \sin^2 B_f + OX^2 - R_a{}^2/2c\ OX \sin B_f)$.

[Eq. 17.9]

Repeat with the second triangle, and add.

Part Three

data

18. Class I Octahedral Coordinates

Eight frequencies — 2, 3, 4, 5, 6, 8, 10, 12 — are here combined in three tables. Pick the vertex labels for the frequency you are using out of the proper column at the left, and find the ϕ and θ coordinates at the right.

Each table covers one octahedron face whose corners are 0,0; 0,90; 90,90. For spheres use only the symmetry triangle (Diagram 12.1). For ellipsoids of revolution (one cross-section circular) use the left half of the triangle. When both floorplans and rooflines are non-circular use the entire triangle.

If plans for coordinate rotation should require data on part of the #2 triangle (down-pointing below the equator) simply repeat ϕ values from the #1 triangle and add 90 to θ values.

A peculiar symmetry of the octahedron permits all coordinates to be written in two notations: arctan and decimal degrees. "Arctan 3" (three keystrokes) is quicker to key on the calculator than "71.5650512" (ten keystrokes) and the entered number is identical. With expressions like "arctan $\sqrt{25}/32$" the saving in keystrokes vanishes but the likelihood of error remains less, symbols on the keyboard being less easily confused than digits. But use the decimal forms if you prefer.

Coordinates were generated by an HP-65 program using Equations 12.1 and 12.2.

OCTAHEDRON CLASS I COORDINATES: Frequencies 12, 6, 3

3ν	6ν	12ν	φ		θ	
0,0	0,0	0,0		00.0000000		00.0000000
		1,0		00.0000000	arctan 1/11	5.1944289
		1,1		90.0000000	arctan 1/11	5.1944289
	1,0	2,0		00.0000000	arctan 1/5	11.3099325
		2,1	arctan 1	45.0000000	arctan √2/10	8.0494670
	1,1	2,2		90.0000000	arctan 1/5	11.3099325
		3,0		00.0000000	arctan 1/3	18.4349488
		3,1	arctan 1/2	26.5650512	arctan √5/81	13.9527364
		3,2	arctan 2	63.4349488	arctan √5/81	13.9527364
		3,3		90.0000000	arctan 1/3	18.4349488
1,0	2,0	4,0		00.0000000	arctan 1/2	26.5650512
		4,1	arctan 1/3	18.4349488	arctan √5/32	21.5681294
	2,1	4,2	arctan 1	45.0000000	arctan √1/8	19.4712206
		4,3	arctan 3	71.5650512	arctan √5/32	21.5681294
1,1	2,2	4,4		90.0000000	arctan 1/2	26.5650512
		5,0		00.0000000	arctan 5/7	35.5376778
		5,1	arctan 1/4	14.0362435	arctan √17/49	30.4987270
		5,2	arctan 2/3	33.6900675	arctan √13/49	27.2520337
		5,3	arctan 3/2	56.3099324	arctan √13/49	27.2520337
		5,4	arctan 4	75.9637566	arctan √17/49	30.4987270
		5,5		90.0000000	arctan 5/7	35.5376778
	3,0	6,0		00.0000000	arctan 1	45.0000000
		6,1	arctan 1/5	11.3099325	arctan √13/18	40.3591005
	3,1	6,2	arctan 1/2	26.5650512	arctan √5/9	36.6992252
		6,3	arctan 1	45.0000000	arctan √1/2	35.2643897
	3,2	6,4	arctan 2	63.4349488	arctan √5/9	36.6992252
		6,5	arctan 5	78.6900675	arctan √13/18	40.3591005
	3,3	6,6		90.0000000	arctan 1	45.0000000
		7,0		00.0000000	arctan 7/5	54.4623222
		7,1	arctan 1/6	9.4623222	arctan √37/25	50.5799684
		7,2	arctan 2/5	21.8014095	arctan √29/25	47.1240113
		7,3	arctan 3/4	36.8698977	arctan 1	45.0000000
		7,4	arctan 4/3	53.1301024	arctan 1	45.0000000
		7,5	arctan 5/2	68.1985905	arctan √29/25	47.1240113
		7,6	arctan 6	80.5376778	arctan √37/25	50.5799684
		7,7		90.0000000	arctan 7/5	54.4623222
2,0	4,0	8,0		00.0000000	arctan 2	63.4349488
		8,1	arctan 1/7	8.1301024	arctan √25/8	60.5037915
	4,1	8,2	arctan 1/3	18.4349488	arctan √5/2	57.6884668
		8,3	arctan 3/5	30.9637565	arctan √17/8	55.5500980

(Continued)

3ν	6ν	12ν	φ		θ	
2,1	4,2	8,4	arctan 1	45.0000000	arctan $\sqrt{2}$	54.7356103
		8,5	arctan 5/3	59.0362435	arctan $\sqrt{17/8}$	55.5500980
	4,3	8,6	arctan 3	71.5650512	arctan $\sqrt{5/2}$	57.6884668
		8,7	arctan 7	81.8698977	arctan $\sqrt{25/8}$	60.5037915
2,2	4,4	8,8		90.0000000	arctan 2	63.4349488
		9,0		00.0000000	arctan 3	71.5650512
		9,1	arctan 1/8	7.1250164	arctan $\sqrt{65/9}$	69.5895545
		9,2	arctan 2/7	15.9453959	arctan $\sqrt{53/9}$	67.6043377
		9,3	arctan 1/2	26.5650512	arctan $\sqrt{5}$	65.9051575
		9,4	arctan 4/5	38.6598083	arctan $\sqrt{41/9}$	64.8959097
		9,5	arctan 5/4	51.3401917	arctan $\sqrt{41/9}$	64.8959097
		9,6	arctan 2	63.4349488	arctan $\sqrt{5}$	65.9051575
		9,7	arctan 3.5	74.0546041	arctan $\sqrt{53/9}$	67.6043377
		9,8	arctan 8	82.8749836	arctan $\sqrt{65/9}$	69.5895545
		9,9		90.0000000	arctan 3	71.5650512
	5,0	10,0		00.0000000	arctan 5	78.6900675
		10,1	arctan 1/9	6.3401918	arctan $\sqrt{41/2}$	77.5454236
	5,1	10,2	arctan 1/4	14.0362435	arctan $\sqrt{17}$	76.3669778
		10,3	arctan 3/7	23.1985905	arctan $\sqrt{29/2}$	75.2856464
	5,2	10,4	arctan 2/3	33.6900675	arctan $\sqrt{13}$	74.4986405
		10,5	arctan 1	45.0000000	arctan $\sqrt{25/2}$	74.2068310
	5,3	10,6	arctan 3/2	56.3099325	arctan $\sqrt{13}$	74.4986405
		10,7	arctan 7/3	66.8014095	arctan $\sqrt{29/2}$	75.2856464
	5,4	10,8	arctan 4	75.9637566	arctan $\sqrt{17}$	76.3669778
		10,9	arctan 9	83.6598082	arctan $\sqrt{41/2}$	77.5454236
	5,5	10,10		90.0000000	arctan 5	78.6900675
		11,0		00.0000000	arctan 11	84.8055711
		11,1	arctan 1/10	5.7105931	arctan $\sqrt{101}$	84.3175615
		11,2	arctan 2/9	12.5288077	arctan $\sqrt{85}$	83.8096005
		11,3	arctan 3/8	20.5560452	arctan $\sqrt{73}$	83.3244079
		11,4	arctan 4/7	29.7448813	arctan $\sqrt{65}$	82.9294449
		11,5	arctan 5/6	39.8055711	arctan $\sqrt{61}$	82.7037247
		11,6	arctan 6/5	50.1944289	arctan $\sqrt{61}$	82.7037247
		11,7	arctan 7/4	60.2551187	arctan $\sqrt{65}$	82.9294449
		11,8	arctan 8/3	69.4439548	arctan $\sqrt{73}$	83.3244079
		11,9	arctan 4.5	77.4711923	arctan $\sqrt{85}$	83.8096005
		11,10	arctan 10	84.2894069	arctan $\sqrt{101}$	84.3175615
		11,11		90.0000000	arctan 11	84.8055711
3,0	6,0	12,0		00.0000000		90.0000000
		12,1	arctan 1/11	5.1944289		90.0000000
	6,1	12,2	arctan 1/5	11.3099325		90.0000000
		12,3	arctan 1/3	18.4349488		90.0000000
3,1	6,2	12,4	arctan 1/2	26.5650512		90.0000000
		12,5	arctan 5/7	35.5376778		90.0000000
	6,3	12,6	arctan 1	45.0000000		90.0000000
		12,7	arctan 7/5	54.4623222		90.0000000
3,2	6,4	12,8	arctan 2	63.4349488		90.0000000
		12,9	arctan 3	71.5650512		90.0000000

(Continued)

				φ		θ
	6,5	12,10	arctan 5	78.6900675		90.0000000
		12,11	arctan 11	84.8055711		90.0000000
3,3	6,6	12,12		90.0000000		90.0000000

OCTAHEDRON CLASS I COORDINATES: Frequencies 8, 4, 2

2ν	4ν	8ν	φ		θ	
0,0	0,0	0,0		00.0000000		00.0000000
		1,0		00.0000000	arctan 1/7	8.1301024
		1,1		90.0000000	arctan 1/7	8.1301024
	1,0	2,0		00.0000000	arctan 1/3	18.4349488
		2,1	arctan 1	45.0000000	arctan √1/18	13.2626760
	1,1	2,2		90.0000000	arctan 1/3	18.4349488
		3,0		00.0000000	arctan 3/5	30.9637565
		3,1	arctan 1/2	26.5650512	arctan √1/5	24.0948426
		3,2	arctan 2	63.4349488	arctan √1/5	24.0948426
		3,3		90.0000000	arctan 3/5	30.9637565
1,0	2,0	4,0		00.0000000	arctan 1	45.0000000
		4,1	arctan 1/3	18.4349488	arctan √5/8	38.3288181
	2,1	4,2	arctan 1	45.0000000	arctan √1/2	35.2643897
		4,3	arctan 3	71.5650512	arctan √5/8	38.3288181
1,1	2,2	4,4		90.0000000	arctan 1	45.0000000
		5,0		00.0000000	arctan 5/3	59.0362435
		5,1	arctan 1/4	14.0362435	arctan √17/9	53.9601066
		5,2	arctan 2/3	33.6900675	arctan √13/9	50.2378408
		5,3	arctan 3/2	56.3099325	arctan √13/9	50.2378408
		5,4	arctan 4	75.9637566	arctan √17/9	53.9601066
		5,5		90.0000000	arctan 5/3	59.0362435
	3,0	6,0		00.0000000	arctan 3	71.5650512
		6,1	arctan 1/5	11.3099325	arctan √13/2	68.5832860
	3,1	6,2	arctan 1/2	26.5650512	arctan √5	65.9051575
		6,3	arctan 1	45.0000000	arctan √9/2	64.7605982
	3,2	6,4	arctan 2	63.4349488	arctan √5	65.9051575
		6,5	arctan 5	78.6900675	arctan √13/2	68.5832860
	3,3	6,6		90.0000000	arctan 3	71.5650512
		7,0		00.0000000	arctan 7	81.8698976
		7,1	arctan 1/6	9.4623222	arctan √37	80.6641409
		7,2	arctan 2/5	21.8014095	arctan √29	79.4802651
		7,3	arctan 3/4	36.8698976	arctan 5	78.6900675
		7,4	arctan 4/3	53.1301024	arctan 5	78.6900675
		7,5	arctan 5/2	68.1985905	arctan √29	79.4802651
		7,6	arctan 6	80.5376778	arctan √37	80.6641409
		7,7		90.0000000	arctan 7	81.8698976

2ν	4ν	8ν	φ	θ
2,0	4,0	8,0	00.0000000	90.0000000
		8,1	arctan 1/7 8.1301024	90.0000000
	4,1	8,2	arctan 1/3 18.4349488	90.0000000
		8,3	arctan 3/5 30.9637565	90.0000000
2,1	4,2	8,4	arctan 1 45.0000000	90.0000000
		8,5	arctan 5/3 59.0362435	90.0000000
	4,3	8,6	arctan 3 71.5650512	90.0000000
		8,7	arctan 7 81.8698976	90.0000000
2,2	4,4	8,8	90.0000000	90.0000000

OCTAHEDRON CLASS I COORDINATES: Frequencies 10, 5, 2

2ν	5ν	10ν	φ	θ
0,0	0,0	0,0	00.0000000	00.0000000
		1,0	00.0000000	arctan 1/9 6.3401917
		1,1	90.0000000	arctan 1/9 6.3401917
	1,0	2,0	00.0000000	arctan 1/4 14.0362435
		2,1	arctan 1 45.0000000	arctan $\sqrt{1/32}$ 10.0249879
	1,1	2,2	90.0000000	arctan 1/4 14.0362435
		3,0	00.0000000	arctan 3/7 23.1985905
		3,1	arctan 1/2 26.5650512	arctan $\sqrt{5/49}$ 17.7154723
		3,2	arctan 2 63.4349488	arctan $\sqrt{5/49}$ 17.7154723
		3,3	90.0000000	arctan 3/7 23.1985905
	2,0	4,0	00.0000000	arctan 2/3 33.6900675
		4,1	arctan 1/3 18.4349488	arctan $\sqrt{5/18}$ 27.7913057
	2,1	4,2	arctan 1 45.0000000	arctan $\sqrt{2/9}$ 25.2394018
		4,3	arctan 3 71.5650512	arctan $\sqrt{5/18}$ 27.7913057
	2,2	4,4	90.0000000	arctan 2/3 33.6900675
1,0		5,0	00.0000000	arctan 1 45.0000000
		5,1	arctan 1/4 14.0362435	arctan $\sqrt{17/25}$ 39.5097123
		5,2	arctan 2/3 33.6900675	arctan $\sqrt{13/25}$ 35.7957599
		5,3	arctan 3/2 56.3099325	arctan $\sqrt{13/25}$ 35.7957599
		5,4	arctan 4 75.9637566	arctan $\sqrt{17/25}$ 39.5097123
1,1		5,5	90.0000000	arctan 1 45.0000000
	3,0	6,0	00.0000000	arctan 3/2 56.3099325
		6,1	arctan 1/5 11.3099325	arctan $\sqrt{13/8}$ 51.8870735
	3,1	6,2	arctan 1/2 26.5650512	arctan $\sqrt{5/4}$ 48.1896851
		6,3	arctan 1 45.0000000	arctan $\sqrt{9/8}$ 46.6861434
	3,2	6,4	arctan 2 63.4349488	arctan $\sqrt{5/4}$ 48.1896851
		6,5	arctan 5 78.6900675	arctan $\sqrt{13/8}$ 51.8870735
	3,3	6,6	90.0000000	arctan 3/2 56.3099325
		7,0	00.0000000	arctan 7/3 66.8014095
		7,1	arctan 1/6 9.4623222	arctan $\sqrt{37/9}$ 63.7476249

(Continued)

2ν	5ν	10ν	ϕ		θ	
		7,2	arctan 2/5	21.8014095	arctan $\sqrt{29/9}$	60.8784319
		7,3	arctan 3/4	36.8698977	arctan $\sqrt{25/9}$	59.0362435
		7,4	arctan 4/3	53.1301024	arctan $\sqrt{25/9}$	59.0362435
		7,5	arctan 5/2	68.1985905	arctan $\sqrt{29/9}$	60.8784319
		7,6	arctan 6	80.5376778	arctan $\sqrt{37/9}$	63.7476249
		7,7		90.0000000	arctan 7/3	66.8014095
	4,0	8,0		00.0000000	arctan 4	75.9637566
		8,1	arctan 1/7	8.1301024	arctan $\sqrt{25/2}$	74.2068310
	4,1	8,2	arctan 1/3	18.4349488	arctan $\sqrt{10}$	72.4515994
		8,3	arctan 3/5	30.9637565	arctan $\sqrt{17/2}$	71.0681768
	4,2	8,4	arctan 1	45.0000000	arctan $\sqrt{8}$	70.5287793
		8,5	arctan 5/3	59.0362435	arctan $\sqrt{17/2}$	71.0681768
	4,3	8,6	arctan 3	71.5650512	arctan $\sqrt{10}$	72.4515994
		8,7	arctan 7	81.8698976	arctan $\sqrt{25/2}$	74.2068310
		8,8		90.0000000	arctan 4	75.9637566
		9,0		00.0000000	arctan 9	83.6598082
		9,1	arctan 1/8	7.1250164	arctan $\sqrt{65}$	82.9294449
		9,2	arctan 2/7	15.9453959	arctan $\sqrt{53}$	82.1787645
		9,3	arctan 1/2	26.5650512	arctan $\sqrt{45}$	81.5212868
		9,4	arctan 4/5	38.6598083	arctan $\sqrt{41}$	81.1236049
		9,5	arctan 5/4	51.3401917	arctan $\sqrt{41}$	81.1236049
		9,6	arctan 2	63.4349488	arctan $\sqrt{45}$	81.5212868
		9,7	arctan 7/2	74.0546041	arctan $\sqrt{53}$	82.1787645
		9,8	arctan 8	82.8749836	arctan $\sqrt{65}$	82.9294449
		9,9		90.0000000	arctan 9	83.6598082
2,0	5,0	10,0		00.0000000		90.0000000
		10,1	arctan 1/9	6.3401918		90.0000000
	5,1	10,2	arctan 1/4	14.0362435		90.0000000
		10,3	arctan 3/7	23.1985905		90.0000000
	5,2	10,4	arctan 2/3	33.6900675		90.0000000
2,1		10,5	arctan 1	45.0000000		90.0000000
	5,3	10,6	arctan 3/2	56.3099325		90.0000000
		10,7	arctan 7/3	66.8014095		90.0000000
	5,4	10,8	arctan 4	75.9637566		90.0000000
		10,9	arctan 9	83.6598082		90.0000000
2,2	5,5	10,10		90.0000000		90.0000000

19. Class II Method 1 Octahedral Coordinates

Two tables provide Class II Method 1 information for the octa breakdown in the following array of frequencies: 16, 12, 8, 6, 4. (The reader will remember that all Class II frequencies are even.)

For the arctan notation see the introductory page for Class I octa coordinates.

To simplify elliptical breakdowns, data for an area equivalent to the entire octa face are given. This consists of the Class II triangle, its downward-pointing reflection below it, and portions to the left and right of the latter: see Diagram 19.1. For spheres use only the symmetry triangle (Diagrams 12.4 and 19.2). For ellipsoids of revolution (one cross-section circular) use the left half of the face. When both roofline and floorplan are non-circular use the entire face.

Class II triangle coordinates were obtained by an HP-65 program using Equations 12.1 and 12.3. A second program then obtained the remaining points by successive rotations using Equations 14.1, 14.2.

19.1

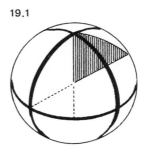

Spherical Octa face
and Class II triangle.

Sphere covered with
Class II triangles.

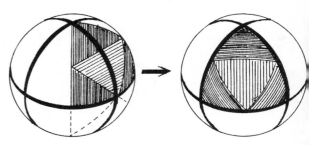

Two Class II triangles
and halves of two others . . . . rotated 45° = Octa face.

19.2

12ν Class II
Octa face and
vertex labels.
(Symmetry
triangle
shaded)

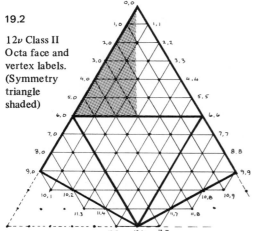

OCTAHEDRON CLASS II COORDINATES: Frequencies 16, 8, 4

4ν	8ν	16ν	ϕ		θ	
0,0	0,0	0,0		00.0000000		00.000000
		1,0		00.0000000	arctan $\sqrt{2/225}$	5.3859771
		1,1		90.0000000	arctan $\sqrt{2/225}$	5.3859771
	1,0	2,0		00.0000000	arctan $\sqrt{2/49}$	11.4217537
		2,1	arctan 1	45.0000000	arctan $\sqrt{1/49}$	8.1301024
	1,1	2,2		90.0000000	arctan $\sqrt{2/49}$	11.4217537
		3,0		00.0000000	arctan $\sqrt{18/169}$	18.0744548
		3,1	arctan 1/2	26.5650512	arctan $\sqrt{10/169}$	13.6717877
		3,2	arctan 2	63.4349488	arctan $\sqrt{10/169}$	13.6717877
		3,3		90.0000000	arctan $\sqrt{18/169}$	18.0744548
1,0	2,0	4,0		00.0000000	arctan $\sqrt{2/9}$	25.2394018
		4,1	arctan 1/3	18.4349488	arctan $\sqrt{5/36}$	20.4393176
	2,1	4,2	arctan 1	45.0000000	arctan $1/3$	18.4349488
		4,3	arctan 3	71.5650512	arctan $\sqrt{5/36}$	20.4393176
1,1	2,2	4,4		90.0000000	arctan $\sqrt{2/9}$	25.2394018
		5,0		00.0000000	arctan $\sqrt{50/121}$	32.7338966
		5,1	arctan 1/4	14.0362435	arctan $\sqrt{34/121}$	27.9274604
		5,2	arctan 2/3	33.6900675	arctan $\sqrt{26/121}$	24.8699503
		5,3	arctan 3/2	56.3099325	arctan $\sqrt{26/121}$	24.8699503
		5,4	arctan 4	75.9637566	arctan $\sqrt{34/121}$	27.9274604
		5,5		90.0000000	arctan $\sqrt{50/121}$	32.7338966
	3,0	6,0		00.0000000	arctan $\sqrt{18/25}$	40.3155422
		6,1	arctan 1/5	11.3099325	arctan $\sqrt{13/25}$	35.7957599
	3,1	6,2	arctan 1/2	26.5650512	arctan $\sqrt{2/5}$	32.3115332
		6,3	arctan 1	45.0000000	arctan $3/5$	30.9637565
	3,2	6,4	arctan 2	63.4349488	arctan $\sqrt{2/5}$	32.3115332
		6,5	arctan 5	78.6900675	arctan $\sqrt{13/25}$	35.7957599
	3,3	6,6		90.0000000	arctan $\sqrt{18/25}$	40.3155422
		7,0		00.0000000	arctan $\sqrt{98/81}$	47.7248561
		7,1	arctan 1/6	9.4623222	arctan $\sqrt{74/81}$	43.7057841
		7,2	arctan 2/5	21.8014095	arctan $\sqrt{58/81}$	40.2377996
		7,3	arctan 3/4	36.8698976	arctan $\sqrt{50/81}$	38.1558065
		7,4	arctan 4/3	53.1301024	arctan $\sqrt{50/81}$	38.1558065
		7,5	arctan 5/2	68.1985905	arctan $\sqrt{58/81}$	40.2377996
		7,6	arctan 6	80.5376778	arctan $\sqrt{74/81}$	43.7057841
		7,7		90.0000000	arctan $\sqrt{98/81}$	47.7248561
2,0	4,0	8,0		00.0000000	arctan $\sqrt{2}$	54.7356103
		8,1	arctan 1/7	8.1301024	arctan $5/4$	51.3401917
	4,1	8,2	arctan 1/3	18.4349488	arctan $\sqrt{5/4}$	48.1896851
		8,3	arctan 3/5	30.9637565	arctan $\sqrt{17/16}$	45.8682508
2,1	4,2	8,4	arctan 1	45.0000000	arctan 1	45.0000000

4ν	8ν	16ν	ϕ		θ	
		8,5	arctan 5/3	59.0362435	arctan $\sqrt{17/16}$	45.8682508
	4,3	8,6	arctan 3	71.5650512	arctan $\sqrt{5}/4$	48.1896851
		8,7	arctan 7	81.8698976	arctan 5/4	51.3401917
2,2	4,4	8,8		90.0000000	arctan $\sqrt{2}$	54.7356103
		9,0		00.0000000	arctan $\sqrt{32/9}$	62.0616473
		9,1	arctan 1/8	7.1250163	arctan $\sqrt{130/49}$	58.4525461
		9,2	arctan 2/7	15.9453959	arctan $\sqrt{106/49}$	55.7882119
		9,3	arctan 1/2	26.5650512	arctan $\sqrt{90/49}$	53.5776902
		9,4	arctan 4/5	38.6598083	arctan $\sqrt{82/49}$	52.2952350
		9,5	arctan 5/4	51.3401917	arctan $\sqrt{82/49}$	52.2952350
		9,6	arctan 2	63.4349488	arctan $\sqrt{90/69}$	53.5776902
		9,7	arctan 7/2	74.0546041	arctan $\sqrt{106/49}$	55.7882119
		9,8	arctan 8	82.8749836	arctan $\sqrt{130/49}$	58.4525461
		9,9		90.0000000	arctan $\sqrt{32/9}$	62.0616473
	5,0	10,0		00.0000000	arctan $\sqrt{8}$	70.5287793
		10,1	arctan 1/8	7.1250163	arctan $\sqrt{26/5}$	66.3211550
	5,1	10,2	arctan 1/4	14.0362435	arctan $\sqrt{34/9}$	62.7743720
		10,3	arctan 3/7	23.1985905	arctan $\sqrt{29/9}$	60.8784319
	5,2	10,4	arctan 2/3	33.6900675	arctan $\sqrt{26/9}$	59.5296405
		10,5	arctan 1	45.0000000	arctan $\sqrt{25/9}$	59.0362435
	5,3	10,6	arctan 3/2	56.3099325	arctan $\sqrt{26/9}$	59.5296405
		10,7	arctan 7/3	66.8014095	arctan $\sqrt{29/9}$	60.8784319
	5,4	10,8	arctan 4	75.9637566	arctan $\sqrt{34/9}$	62.7743720
		10,9	arctan 8	82.8749836	arctan $\sqrt{26/5}$	66.3211550
	5,5	10,10		90.0000000	arctan $\sqrt{8}$	70.5287793
		11,0		00.0000000	arctan $\sqrt{32}$	79.9750121
		11,1	arctan 1/8	7.1250163	arctan $\sqrt{130/9}$	75.2586068
		11,2	arctan 1/4	14.0362435	arctan $\sqrt{17/2}$	71.0681768
		11,3	arctan 3/8	20.5560452	arctan $\sqrt{146/25}$	67.5200842
		11,4	arctan 4/7	29.7448813	arctan $\sqrt{26/5}$	66.3211550
		11,5	arctan 5/6	39.8055711	arctan $\sqrt{122/25}$	65.6447487
		11,6	arctan 6/5	50.1944289	arctan $\sqrt{122/25}$	65.6447487
		11,7	arctan 7/4	60.2551187	arctan $\sqrt{26/5}$	66.3211550
		11,8	arctan 8/3	69.4439548	arctan $\sqrt{146/25}$	67.5200842
		11,9	arctan 4	75.9637566	arctan $\sqrt{17/2}$	71.0681768
		11,10	arctan 8	82.8749836	arctan $\sqrt{130/9}$	75.2586068
		11,11		90.0000000	arctan $\sqrt{32}$	79.9750121
3,0	6,0	12,0		00.0000000		90.0000000
		12,1	arctan 1/8	7.1250163	arctan $\sqrt{130}$	84.9876536
	6,1	12,2	arctan 1/4	14.0362435	arctan $\sqrt{34}$	80.2685245
		12,3	arctan 3/8	20.5560452	arctan $\sqrt{146/9}$	76.0564504
3,1	6,2	12,4	arctan 1/2	26.5650512	arctan $\sqrt{10}$	72.4515994
		12,5	arctan 5/7	35.5376778	arctan $\sqrt{37/4}$	71.7992398
	6,3	12,6	arctan 1	45.0000000	arctan 3	71.5650512
		12,7	arctan 7/5	54.4623222	arctan $\sqrt{37/4}$	71.7992398
3,2	6,4	12,8	arctan 2	63.4349488	arctan $\sqrt{10}$	72.4515994
		12,9	arctan 8/3	69.4439548	arctan $\sqrt{146/9}$	76.0564504
	6,5	12,10	arctan 4	75.9637566	arctan $\sqrt{34}$	80.2685245

4ν	8ν	16ν	ϕ		θ	
		12,11	arctan 8	82.8749836	arctan $\sqrt{130}$	84.9876536
3,3	6,6	12,12		90.0000000		90.0000000
		13,1	arctan 1/8	7.1250163	arctan $-\sqrt{130}$	95.0123464
		13,2	arctan 1/4	14.0362435		90.0000000
		13,3	arctan 3/8	20.5560452	arctan $\sqrt{146}$	85.2689493
		13,4	arctan 1/2	26.5650512	arctan $\sqrt{40}$	81.0151231
		13,5	arctan 5/8	32.0053832	arctan $\sqrt{178/9}$	77.3272785
		13,6	arctan 6/7	40.6012947	arctan $\sqrt{170/9}$	77.0423654
		13,7	arctan 7/6	49.3987054	arctan $\sqrt{170/9}$	77.0423654
		13,8	arctan 8/5	57.9946168	arctan $\sqrt{178/9}$	77.3272785
		13,9	arctan 2	63.4349488	arctan $\sqrt{40}$	81.0151231
		13,10	arctan 8/3	69.4439548	arctan $\sqrt{146}$	85.2689493
		13,11	arctan 4	75.9637566		90.0000000
		13,12	arctan 8	82.8749836	arctan $-\sqrt{130}$	95.0123464
		14,3	arctan 3/8	20.5560452	arctan $-\sqrt{146}$	94.7310506
	7,2	14,4	arctan 1/2	26.5650512		90.0000000
		14,5	arctan 5/8	32.0053832	arctan $\sqrt{178}$	85.7135149
	7,3	14,6	arctan 3/4	36.8698976	arctan $\sqrt{50}$	81.9505330
		14,7	arctan 1	45.0000000	arctan 7	81.8698976
	7,4	14,8	arctan 4/3	53.1301024	arctan $\sqrt{50}$	81.9505330
		14,9	arctan 8/5	57.9946168	arctan $\sqrt{178}$	85.7135149
	7,5	14,10	arctan 2	63.4349488		90.0000000
		14,11	arctan 8/3	69.4439548	arctan $-\sqrt{146}$	94.7310506
		15,5	arctan 5/8	32.0053832	arctan $-\sqrt{178}$	94.2864851
		15,6	arctan 3/4	36.8698976		90.0000000
		15,7	arctan 7/8	41.1859252	arctan $\sqrt{226}$	86.1943478
		15,8	arctan 8/7	48.8140748	arctan $\sqrt{226}$	86.1943478
		15,9	arctan 4/3	53.1301024		90.0000000
		15,10	arctan 8/5	57.9946168	arctan $-\sqrt{178}$	94.2864851
		16,7	arctan 7/8	41.1859252	arctan $-\sqrt{226}$	93.8056521
4,2	8,4	16,8	arctan 1	45.0000000		90.0000000
		16,9	arctan 8/7	48.8140748	arctan $-\sqrt{226}$	93.8056521

OCTAHEDRON CLASS II COORDINATES: Frequencies 12, 6

6ν	12ν	ϕ		θ	
0,0	0,0		00.0000000		00.0000000
	1,0		00.0000000	arctan $\sqrt{2/121}$	7.3260370
	1,1		90.0000000	arctan $\sqrt{2/121}$	7.3260370
1,0	2,0		00.0000000	arctan $\sqrt{2/25}$	15.7931691
	2,1	arctan 1	45.0000000	arctan $1/5$	11.3099325
1,1	2,2		90.0000000	arctan $\sqrt{2/25}$	15.7931691
	3,0		00.0000000	arctan $\sqrt{2/9}$	25.2394018
	3,1	arctan 1/2	26.5650512	arctan $\sqrt{10/81}$	19.3596487
	3,2	arctan 2	63.4349488	arctan $\sqrt{10/81}$	19.3596487
	3,3		90.0000000	arctan $\sqrt{2/9}$	25.2394018
2,0	4,0		00.0000000	arctan $\sqrt{1/2}$	35.2643897
	4,1	arctan 1/3	18.4349488	arctan $\sqrt{5/16}$	29.2059322
2,1	4,2	arctan 1	45.0000000	arctan $1/2$	26.5650512
	4,3	arctan 3	71.5650512	arctan $\sqrt{5/16}$	29.2059322
2,2	4,4		90.0000000	arctan $\sqrt{1/2}$	35.2643897
	5,0		00.0000000	arctan $\sqrt{50/49}$	45.2893775
	5,1	arctan 1/4	14.0362435	arctan $\sqrt{34/49}$	39.7940655
	5,2	arctan 2/3	33.6900675	arctan $\sqrt{26/49}$	36.0707687
	5,3	arctan 3/2	56.3099325	arctan $\sqrt{26/49}$	36.0707687
	5,4	arctan 4	75.9637566	arctan $\sqrt{34/49}$	39.7940655
	5,5		90.0000000	arctan $\sqrt{50/49}$	45.2893775
3,0	6,0		00.0000000	arctan $\sqrt{2}$	54.7356103
	6,1	arctan 1/5	11.3099325	arctan $\sqrt{13/9}$	50.2378408
3,1	6,2	arctan 1/2	26.5650512	arctan $\sqrt{10/9}$	46.5084806
	6,3	arctan 1	45.0000000	arctan 1	45.0000000
3,2	6,4	arctan 2	63.4349488	arctan $\sqrt{10/9}$	46.5084806
	6,5	arctan 5	78.6900675	arctan $\sqrt{13/9}$	50.2378408
3,3	6,6		90.0000000	arctan $\sqrt{2}$	54.7356103
	7,0		00.0000000	arctan $\sqrt{9/2}$	64.7605982
	7,1	arctan 1/6	9.4623222	arctan $\sqrt{74/25}$	59.8332100
	7,2	arctan 2/5	21.8014095	arctan $\sqrt{58/25}$	56.7138133
	7,3	arctan 3/4	36.8698976	arctan $\sqrt{2}$	54.7356103
	7,4	arctan 4/3	53.1301024	arctan $\sqrt{2}$	54.7356103
	7,5	arctan 5/2	68.1985905	arctan $\sqrt{58/25}$	56.7138133
	7,6	arctan 6	80.5376778	arctan $\sqrt{74/25}$	59.8332100
	7,7		90.0000000	arctan $\sqrt{9/2}$	64.7605982
4,0	8,0		00.0000000	arctan $\sqrt{18}$	76.7373240
	8,1	arctan 1/6	9.4623222	arctan $\sqrt{74/9}$	70.7741443
4,1	8,2	arctan 1/3	18.4349488	arctan $\sqrt{5}$	65.9051575
	8,3	arctan 3/5	30.9637565	arctan $\sqrt{17/4}$	64.1233099
4,2	8,4	arctan 1	45.0000000	arctan 2	63.4349488
	8,5	arctan 5/3	59.0362435	arctan $\sqrt{17/4}$	64.1233099

(Continued)

6ν	12ν	φ		θ	
4,3	8,6	arctan 3	71.5650512	arctan √5	65.9051575
	8,7	arctan 6	80.5376778	arctan √74/9	70.7741443
4,4	8,8		90.0000000	arctan √18	76.7373240
	9,0		00.0000000		90.0000000
	9,1	arctan 1/6	9.4623222	arctan √74	83.3692622
	9,2	arctan 1/3	18.4349488	arctan √20	77.3956173
	9,3	arctan 1/2	26.5650512	arctan √10	72.4515994
	9,4	arctan 4/5	38.6598083	arctan √82/9	71.6702463
	9,5	arctan 5/4	51.3401917	arctan √82/9	71.6702463
	9,6	arctan 2	63.4349488	arctan √10	72.4515994
	9,7	arctan 3	71.5650512	arctan √20	77.3956173
	9,8	arctan 6	80.5376778	arctan √74	83.3692622
	9,9		90.0000000		90.0000000
	10,1	arctan 1/6	9.4623222	arctan −√74	96.6307378
5,1	10,2	arctan 1/3	18.4349488		90.0000000
	10,3	arctan 1/2	26.5650512	arctan √90	83.9827152
5,2	10,4	arctan 2/3	33.6900675	arctan √26	78.9041967
	10,5	arctan 1	45.0000000	arctan 5	78.6900675
5,3	10,6	arctan 3/2	56.3099325	arctan √26	78.9041967
	10,7	arctan 2	63.4349488	arctan √90	83.9827152
5,4	10,8	arctan 3	71.5650512		90.0000000
	10,9	arctan 6	80.5376778	arctan −√74	96.6307378
	11,3	arctan 1/2	26.5650512	arctan −√90	96.0172849
	11,4	arctan 2/3	33.6900675		90.0000000
	11,5	arctan 5/6	39.8055711	arctan √122	84.8267875
	11,6	arctan 6/5	50.1944289	arctan √122	84.8267875
	11,7	arctan 3/2	56.3099325		90.0000000
	11,8	arctan 2	63.4349488	arctan −√90	96.0172849
6,6	12,5	arctan 5/6	39.8055711	arctan −√122	95.1732124
	12,6	arctan 1	45.0000000		90.0000000
	12,7	arctan 6/5	50.1944289	arctan −√122	95.1732124

20. Class I Method 1 Icosahedral Coordinates

Data are given for Class I icosa breakdowns of the following frequencies: 10, 8, 6, 5, 4, 3, 2. Pick the vertex labels for the frequency you are using out of the proper column at the left, and read ϕ and θ values from the two columns at the right.

Each table covers *two* icosa faces: the #1 triangle whose corners are 0,0; 0, arctan 2; 72, arctan 2; and the downward-pointing #2 triangle beneath it whose corners are 0, arctan 2; 72, arctan 2; 36, (180 – arctan 2): see Diagram 12.3. For spheres use only the Symmetry Triangle shown in Diagram 12.2. For ellipsoids of revolution (one cross-section circular) use the left-hand half of both triangles as far down as the proposed truncation plane. For structures with two non-circular cross-sections you need a 90° span of ϕ values as well. Diagram 20.1 shows how these can be extrapolated. But such structures have more tractable symmetries if rotated so that an edge, not a vertex, is at the zenith: see Chapter 15.

An HP-65 program using Equations 12.4 through 12.8 generated the values for Triangle #1. A second program using Equations 14.1, 14.2 then generated the Triangle #2 points by rotation.

20.1

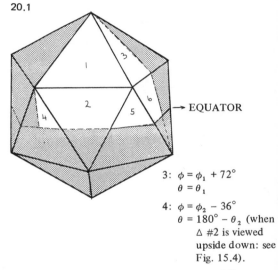

→ EQUATOR

3: $\phi = \phi_1 + 72°$
$\theta = \theta_1$

4: $\phi = \phi_2 - 36°$
$\theta = 180° - \theta_2$ (when $\triangle$ #2 is viewed upside down: see Fig. 15.4).

5: $\phi = \phi_2 + 36°$
$\theta = $ same as $\triangle$ #4.

6: $\phi = 72° + \phi_2$
$\theta = \theta_2$.

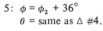

ICOSAHEDRON CLASS I COORDINATES: Frequencies 10, 5

5ν	10ν	φ	θ
0,0	0,0	00.0000000	00.0000000
	1,0	00.0000000	5.4084310
	1,1	72.0000000	5.4084310
1,0	2,0	00.0000000	11.3716668
	2,1	36.0000000	9.2416169
1,1	2,2	72.0000000	11.3716668
	3,0	00.0000000	17.8315873
	3,1	22.3861776	14.9899922
	3,2	49.6138225	14.9899922
	3,3	72.0000000	17.8315873
2,0	4,0	00.0000000	24.6710680
	4,1	16.0353713	21.5724472
2,1	4,2	36.0000000	20.3856407
	4,3	55.9646288	21.5724472
2,2	4,4	72.0000000	24.6710680
	5,0	00.0000000	31.7174744
	5,1	12.4463843	28.6099376
	5,2	27.7323015	26.8052365
	5,3	44.2676986	26.8052365
	5,4	59.5536157	28.6099376
	5,5	72.0000000	31.7174744
3,0	6,0	00.0000000	38.7638808
	6,1	10.1562304	35.8224702
3,1	6,2	22.3861776	33.7590240
	6,3	36.0000000	33.0088450
3,2	6,4	49.6138225	33.7590240
	6,5	61.8437696	35.8224702
3,3	6,6	72.0000000	38.7638808
	7,0	00.0000000	45.6033616
	7,1	8.5725404	42.9496334
	7,2	18.7046980	40.8717741
	7,3	30.0743861	39.7156131
	7,4	41.9256139	39.7156131
	7,5	53.2953021	40.8717741
	7,6	63.4274596	42.9496334
	7,7	72.0000000	45.6033616
4,0	8,0	00.0000000	52.0632820
	8,1	7.4137289	49.7664590
4,1	8,2	16.0353713	47.8353485
	8,3	25.7052785	46.5286650
4,2	8,4	36.0000000	46.0642057

5ν	10ν	φ	θ
4,3	8,5	46.2947215	46.5286650
	8,6	55.9646288	47.8353485
	8,7	64.5862711	49.7664590
4,4	8,8	72.0000000	52.0632820
	9,0	00.0000000	58.0265179
	9,1	6.5297024	56.1087309
	9,2	14.0192855	54.4163133
	9,3	22.3861776	53.1334950
	9,4	31.3846950	52.4366973
	9,5	40.6153050	52.4366973
	9,6	49.6138225	53.1334950
	9,7	57.9807145	54.4163133
	9,8	65.4702976	56.1087309
	9,9	72.0000000	58.0265179
5,0	10,0	00.0000000	63.4349488
	10,1	5.8333887	61.8829712
5,1	10,2	12.4463843	60.4665625
	10,3	19.7952367	59.3117249
5,2	10,4	27.7323015	58.5495950
	10,5	36.0000000	58.2825256
5,3	10,6	44.2676986	58.5495950
	10,7	52.2047633	59.3117249
5,4	10,8	59.5536157	60.4665625
	10,9	66.1666114	61.8829712
5,5	10,10	72.0000000	63.4349488

	(Triangle #2)		
	11,1	3.4292316	67.8508314
	11,2	9.5794832	66.6079548
	11,3	16.4605636	65.5245982
	11,4	23.9792000	64.7124135
	11,5	31.9400304	64.2746575
	11,6	40.0599696	64.2746575
	11,7	48.0207000	64.7124135
	11,8	55.5394364	65.5245982
	11,9	62.4205168	66.6079548
	11,10	68.5707684	67.8508314
6,1	12,2	6.9684505	72.7969533
	12,3	13.3994138	71.8969574
6,2	12,4	20.4898913	71.1600132
	12,5	28.1002356	70.6721118
6,3	12,6	36.0000000	70.5008455
	12,7	43.8997644	70.6721118
6,4	12,8	51.5101087	71.1600132
	12,9	58.6005862	71.8969574
6,5	12,10	65.0315495	72.7969533

(Continued)

5ν	10ν	φ	θ
	13,3	10.5948977	78.2204432
	13,4	17.2607131	77.6677097
	13,5	24.4947652	77.2528359
	13,6	32.1184790	77.0290522
	13,7	39.8815210	77.0290522
	13,8	47.5052349	77.2528359
	13,9	54.7392869	77.6677097
	13,10	61.4051023	78.2204432
7,2	14,4	14.2819224	84.0101625
	14,5	21.1287296	83.7703158
7,3	14,6	28.4369994	83.6117970
	14,7	36.0000000	83.5562062
7,4	14,8	43.5630006	83.6117970
	14,9	50.8712704	83.7703158
7,5	14,10	57.7180776	84.0101625
	15,5	18.0000000	90.0000000
	15,6	24.9684505	90.0000000
	15,7	32.2819224	90.0000000
	15,8	39.7180776	90.0000000
	15,9	47.0315495	90.0000000

5ν	10ν	φ	θ
	15,10	54.0000000	90.0000000
8,3	16,6	21.7180776	95.9898376
	16,7	28.7463723	96.1303345
8,4	16,8	36.0000000	96.1794105
	16,9	43.2536277	96.1303345
8,5	16,10	50.2819224	95.9898376
	17,7	25.4051023	101.7795569
	17,8	32.4321744	101.9554615
	17,9	39.5678256	101.9554615
	17,10	46.5948977	101.7795569
9,4	18,8	29.0315495	107.2030468
	18,9	36.0000000	107.3234344
9,5	18,10	42.9684505	107.2030468
	19,9	32.5707684	112.1491687
	19,10	39.4292316	112.1491687
10,5	20,10	36.0000000	116.5650512

ICOSAHEDRON CLASS I COORDINATES: Frequencies 8, 4, 2

2ν	4ν	8ν	φ	θ
0,0	0,0	0,0	00.0000000	00.0000000
		1,0	00.0000000	6.8485493
		1,1	72.0000000	6.8485493
	1,0	2,0	00.0000000	14.5454366
		2,1	36.0000000	11.8548751
	1,1	2,2	72.0000000	14.5454366
		3,0	00.0000000	22.9342438
		3,1	22.3861776	19.4026533
		3,2	49.6138225	19.4026533
		3,3	72.0000000	22.9342438
1,0	2,0	4,0	00.0000000	31.7174744
		4,1	16.0353713	28.0116135
	2,1	4,2	36.0000000	26.5650512
		4,3	55.9646288	28.0116135
1,1	2,2	4,4	72.0000000	31.7174744
		5,0	00.0000000	40.5007050
		5,1	12.4463843	37.0083770
		5,2	27.7323015	34.9242784
		5,3	44.2676986	34.9242784
		5,4	59.5536157	37.0083770
		5,5	72.0000000	40.5007050
	3,0	6,0	00.0000000	48.8895123
		6,1	10.1562304	45.8486935
	3,1	6,2	22.3861776	43.6469271
		6,3	36.0000000	42.8320885
	3,2	6,4	49.6138225	43.6469271
		6,5	61.8437696	45.8486935
	3,3	6,6	72.0000000	48.8895123
		7,0	00.0000000	56.5863995
		7,1	8.5725404	54.1031325
		7,2	18.7046980	52.0961412
		7,3	30.0743861	50.9544734
		7,4	41.9256139	50.9544734
		7,5	53.2953021	52.0961412
		7,6	63.4274596	54.1031325
		7,7	72.0000000	56.5863995

2ν	4ν	8ν	φ	θ
2,0	4,0	8,0	00.0000000	63.4349488
		8,1	7.4137289	61.5115731
	4,1	8,2	16.0353713	59.8480232
		8,3	25.7052785	58.6968610
2,1	4,2	8,4	36.0000000	58.2825256
		8,5	46.2947215	58.6968610
	4,3	8,6	55.9646288	59.8480232
		8,7	64.5862711	61.5115731
2,2	4,4	8,8	72.0000000	63.4349488
(Triangle #2)				
		9,1	4.3045403	69.0387834
		9,2	12.1989876	67.6108929
		9,3	21.1768230	66.4786457
		9,4	30.9587180	65.8438654
		9,5	41.0412821	65.8438654
		9,6	50.8231770	66.4786457
		9,7	59.8010124	67.6108929
		9,8	67.6954597	69.0387834
	5,1	10,2	8.7723551	75.4545635
		10,3	17.0666553	74.5700194
	5,2	10,4	26.2676986	73.9549430
		10,5	36.0000000	73.7329627
	5,3	10,6	45.7323015	73.9549430
		10,7	54.9333447	74.5700194
	5,4	10,8	63.2276449	75.4545635
		11,3	13.3560643	82.5366786
		11,4	21.9478021	82.1599429
		11,5	31.2307464	81.9488415
		11,6	40.7692536	81.9488415
		11,7	50.0521979	82.1599429
		11,8	58.6439357	82.5366786
3,1	6,2	12,4	18.0000000	90.0000000
		12,5	26.7723551	90.0000000
	6,3	12,6	36.0000000	90.0000000
		12,7	45.2276449	90.0000000
3,2	6,4	12,8	54.0000000	90.0000000

(Continued)

2ν	4ν	8ν	φ	θ
		13,5	22.6439357	97.4633215
		13,6	31.4749784	97.6447358
		13,7	40.5250216	97.6447358
		13,8	49.3560643	97.4633215
	7,3	14,6	27.2276449	104.5454366
		14,7	36.0000000	104.7101762

2ν	4ν	8ν	φ	θ
	7,4	14,8	44.7723551	104.5454366
		15,7	31.6954597	110.9612167
		15,8	40.3045403	110.9612167
4,2	8,4	16,8	36.0000000	116.5650512

ICOSAHEDRON CLASS I COORDINATES: Frequencies 6, 3

3ν	6ν	φ	θ
0,0	0,0	00.0000000	00.0000000
	1,0	00.0000000	9.3247035
	1,0	72.0000000	9.3247035
1,0	2,0	00.0000000	20.0767513
	2,1	36.0000000	16.4722107
1,1	2,2	72.0000000	20.0767513
	3,0	00.0000000	31.7174744
	3,1	22.3861776	27.2237351
	3,2	49.6138225	27.2237351
	3,3	72.0000000	31.7174744
2,0	4,0	00.0000000	43.3581976
	4,1	16.0353713	39.1034177
2,1	4,2	36.0000000	37.3773682
	4,3	55.9646288	39.1034177
2,2	4,4	72.0000000	43.3581976
	5,0	00.0000000	54.1102453
	5,1	12.4463843	50.6513527
	5,2	27.7323015	48.4869490
	5,3	44.2676986	48.4869490
	5,4	59.5536157	50.6513527
	5,5	72.0000000	54.1102453
3,0	6,0	00.0000000	63.4349488
	6,1	10.1562304	60.9162275
3,1	6,2	22.3861776	59.0080312
	6,3	36.0000000	58.2825256
3,2	6,4	49.6138225	59.0080312
	6,5	61.8437696	60.9162276
3,3	6,6	72.0000000	63.4349488

3ν	6ν	φ	θ
	7,1	5.7778072	71.0914502
	7,2	16.7344818	69.4828700
	7,3	29.3546281	68.5073635
	7,4	42.6453719	68.5073635
	7,5	55.2655182	69.4828700
	7,6	66.2221928	71.0914502
4,1	8,2	11.8185857	80.1168545
	8,3	23.3461239	79.4443283
4,2	8,4	36.0000000	79.1876831
	8,5	48.6538761	79.4443283
4,3	8,6	60.1814143	80.1168545
	9,3	18.0000000	90.0000000
	9,4	29.8185857	90.0000000
	9,5	42.1814143	90.0000000
	9,6	54.0000000	90.0000000
5,2	10,4	24.1814143	99.8831455
	10,5	36.0000000	100.0928406
5,3	10,6	47.8185858	99.8831455
	11,5	30.2221928	108.9085499
	11,6	41.7778072	108.9085499
6,3	12,6	36.0000000	116.5650512

← *Triangle #1*

↑ *Triangle #2*

21. Class II Method 1 Icosahedral Coordinates

Two tables give Class II Method 1 information for the icosa breakdown of frequencies 12, 8, 6, 4, 2. (The reader will remember that all Class II frequencies are even.)

The unit of attention is not the icosa face but its Class II triangle (see Chapter 10). Five of these triangles are covered in the tables: see Diagram 12.5 and the explanation in Chapter 12. Ellipsoidal structures can thus be computed as far down as an equatorial truncation plane. For spheres we disregard most of the data provided and use only the left half of the top Class II triangle (Diagram 12.4); this would be, for example, the area bounded by 0,0; 4,0; 4,2 in the 8ν case shown in Diagram 12.5.

An HP-65 program using Equations 12.4, 12.5, 12.6, 12.7, 12.9 generated the Class II triangle data; a second program then obtained the remaining four triangles by successive rotations using Equations 14.1 and 14.2.

ICOSAHEDRON CLASS II COORDINATES: Frequencies 12, 6

6ν	12ν	ϕ	θ		6ν	12ν	ϕ	θ
0,0	0,0	00.0000000	00.0000000			9,0	00.0000000	58.2825256
						9,1	7.1842330	54.8527896
	1,0	00.0000000	5.5165213			9,2	14.7278968	51.9590943
	1,1	72.0000000	5.5165213			9,3	22.3861776	49.6967416
						9,4	31.3846950	48.9816797
1,0	2,0	00.0000000	11.4729759			9,5	40.6153050	48.9816797
	2,1	36.0000000	9.3247035			9,6	49.6138224	49.6967416
1,1	2,2	72.0000000	11.4729759		5,0	10,0	00.0000000	65.5384985
	3,0	00.0000000	17.7818206			10,1	6.6453720	62.0852025
	3,1	22.3861776	14.9473413		5,1	10,2	13.6138224	59.0080313
	3,2	49.6138225	14.9473413			10,3	20.7140829	56.4260598
	3,3	72.0000000	17.7818206		5,2	10,4	27.7323015	54.3940654
2,0	4,0	00.0000000	24.3119302			10,5	36.0000000	54.1102454
	4,1	16.0353713	21.2489135		5,3	10,6	44.2676985	54.3940654
2,1	4,2	36.0000000	20.0767513			11,0	00.0000000	75.5689243
	4,3	55.9646288	21.2489135			11,1	6.1814143	69.2059650
2,2	4,4	72.0000000	24.3119302			11,2	12.6538762	66.0652074
	5,0	00.0000000	30.9008742			11,3	19.2655182	63.2739586
	5,1	12.4463843	27.8434843			11,4	25.8437697	60.9162276
	5,2	27.7323015	26.0719770			11,5	32.2211414	59.0222838
	5,3	44.2676986	26.0719770			11,6	39.7788586	59.0222838
	5,4	59.5536157	27.8434843		6,0	12,0	00.0000000	79.1876831
	5,5	72.0000000	30.9008742			12,1	5.7778072	76.0076786
3,0	6,0	00.0000000	37.3773681		6,1	12,2	11.8185859	72.9167205
	6,1	10.1562304	34.4779740			12,3	18.0000000	70.0353713
3,1	6,2	22.3861776	32.4522840		6,2	12,4	24.1814143	67.4616665
	6,3	36.0000000	31.7174744			12,5	30.2221929	65.2555423
3,2	6,4	49.6138225	32.4522840		6,3	12,6	36.0000000	63.4349488
	6,5	61.8437696	34.4779740			13,0	00.0000000	85.6641771
3,3	6,6	72.0000000	37.3773681			13,1	5.7778072	82.5457713
	7,0	00.0000000	43.9961269			13,2	11.8185859	79.4116562
	7,1	8.5725404	40.9443707			13,3	18.0000000	76.3861775
	7,2	18.7046980	38.8872075			13,4	24.1814143	73.5844349
	7,3	30.0743861	37.7470129			13,5	30.2221929	71.0914502
	7,4	41.9256139	37.7470129			13,6	36.0000000	68.9514701
	7,5	53.2953020	38.8872076		7,0	14,0	00.0000000	92.2531211
	7,6	63.4274595	40.9443707			14,1	5.7778072	89.2841783
4,0	8,0	00.0000000	51.0265527		7,1	14,2	11.8185857	86.1926965
	8,1	7.8176058	47.7365997			14,3	18.0000000	83.0956449
4,1	8,2	16.0353713	45.1328359		7,2	14,4	24.1814143	80.1168545
	8,3	25.7052785	43.8225702			14,5	30.2221929	77.3633902
4,2	8,4	36.0000000	43.3581976		7,3	14,6	36.0000000	74.9079247
	8,5	46.2947215	43.8225702				*(Continued)*	
4,3	8,6	55.9646287	45.1328359					

21. Class II Method 1 Icosahedral Coordinates

Two tables give Class II Method 1 information for the icosa breakdown of frequencies 12, 8, 6, 4, 2. (The reader will remember that all Class II frequencies are even.)

The unit of attention is not the icosa face but its Class II triangle (see Chapter 10). Five of these triangles are covered in the tables: see Diagram 12.5 and the explanation in Chapter 12. Ellipsoidal structures can thus be computed as far down as an equatorial truncation plane. For spheres we disregard most of the data provided and use only the left half of the top Class II triangle (Diagram 12.4); this would be, for example, the area bounded by 0,0; 4,0; 4,2 in the 8ν case shown in Diagram 12.5.

An HP-65 program using Equations 12.4, 12.5, 12.6, 12.7, 12.9 generated the Class II triangle data; a second program then obtained the remaining four triangles by successive rotations using Equations 14.1 and 14.2.

ICOSAHEDRON CLASS II COORDINATES: Frequencies 12, 6

6ν	12ν	φ	θ
0,0	0,0	00.0000000	00.0000000
	1,0	00.0000000	5.5165213
	1,1	72.0000000	5.5165213
1,0	2,0	00.0000000	11.4729759
	2,1	36.0000000	9.3247035
1,1	2,2	72.0000000	11.4729759
	3,0	00.0000000	17.8818206
	3,1	22.3861776	14.9473413
	3,2	49.6138225	14.9473413
	3,3	72.0000000	17.7818206
2,0	4,0	00.0000000	24.3119302
	4,1	16.0353713	21.2489135
2,1	4,2	36.0000000	20.0767513
	4,3	55.9646288	21.2489135
2,2	4,4	72.0000000	24.3119302
	5,0	00.0000000	30.9008742
	5,1	12.4463843	27.8434843
	5,2	27.7323015	26.0719770
	5,3	44.2676986	26.0719770
	5,4	59.5536157	27.8434843
	5,5	72.0000000	30.9008742
3,0	6,0	00.0000000	37.3773681
	6,1	10.1562304	34.4779740
3,1	6,2	22.3861776	32.4522840
	6,3	36.0000000	31.7174744
3,2	6,4	49.6138225	32.4522840
	6,5	61.8437696	34.4779740
3,3	6,6	72.0000000	37.3773681
	7,0	00.0000000	43.9961269
	7,1	8.5725404	40.9443707
	7,2	18.7046980	38.8872075
	7,3	30.0743861	37.7470129
	7,4	41.9256139	37.7470129
	7,5	53.2953020	38.8872076
	7,6	63.4274595	40.9443707
4,0	8,0	00.0000000	51.0265527
	8,1	7.8176058	47.7365997
4,1	8,2	16.0353713	45.1328359
	8,3	25.7052785	43.8225702
4,2	8,4	36.0000000	43.3581976
	8,5	46.2947215	43.8225702
4,3	8,6	55.9646287	45.1328359

6ν	12ν	φ	θ
	9,0	00.0000000	58.2825256
	9,1	7.1842330	54.8527896
	9,2	14.7278968	51.9590943
	9,3	22.3861776	49.6967416
	9,4	31.3846950	48.9816797
	9,5	40.6153050	48.9816797
	9,6	49.6138224	49.6967416
5,0	10,0	00.0000000	65.5384985
	10,1	6.6453720	62.0852025
5,1	10,2	13.6138224	59.0080313
	10,3	20.7140829	56.4260598
5,2	10,4	27.7323015	54.3940654
	10,5	36.0000000	54.1102454
5,3	10,6	44.2676985	54.3940654
	11,0	00.0000000	75.5689243
	11,1	6.1814143	69.2059650
	11,2	12.6538762	66.0652074
	11,3	19.2655182	63.2739586
	11,4	25.8437697	60.9162276
	11,5	32.2211414	59.0222838
	11,6	39.7788586	59.0222838
6,0	12,0	00.0000000	79.1876831
	12,1	5.7778072	76.0076786
6,1	12,2	11.8185859	72.9167205
	12,3	18.0000000	70.0353713
6,2	12,4	24.1814143	67.4616665
	12,5	30.2221929	65.2555423
6,3	12,6	36.0000000	63.4349488
	13,0	00.0000000	85.6641771
	13,1	5.7778072	82.5457713
	13,2	11.8185859	79.4116562
	13,3	18.0000000	76.3861775
	13,4	24.1814143	73.5844349
	13,5	30.2221929	71.0914502
	13,6	36.0000000	68.9514701
7,0	14,0	00.0000000	92.2531211
	14,1	5.7778072	89.2841783
7,1	14,2	11.8185857	86.1926965
	14,3	18.0000000	83.0956449
7,2	14,4	24.1814143	80.1168545
	14,5	30.2221929	77.3633902
7,3	14,6	36.0000000	74.9079247

(Continued)

(Continued)

6ν	12ν	φ	θ
	15,0	00.0000000	98.7832307
	15,1	5.7778072	96.0423546
	15,2	11.8185857	93.0817398
	15,3	18.0000000	90.0000000
	15,4	24.1814143	86.9182604
	15,5	30.2221929	83.9576455
	15,6	36.0000000	81.2167695
8,0	16,0	00.0000000	105.0920754
	16,1	5.7778072	102.6366099
8,1	16,2	11.8185858	99.8831455
	16,3	18.0000000	96.9043551
8,2	16,4	24.1814143	93.8073037
	16,5	30.2221929	90.7158218
8,3	16,6	36.0000000	87.7468790

6ν	12ν	φ	θ
	17,0	00.0000000	111.0485300
	17,1	5.7778072	108.9085499
	17,2	11.8185857	106.4155651
	17,3	18.0000000	103.6138226
	17,4	24.1814143	100.5883439
	17,5	30.2221929	97.4542289
	17,6	36.0000000	94.3358230
9,0	18,0	00.0000000	116.5650512
	18,1	5.7778072	114.7444578
9,1	18,2	11.8185857	112.5383336
	18,3	18.0000000	109.9646288
9,2	18,4	24.1814143	107.0832796
	18,5	30.2221929	103.9923215
9,3	18,6	36.0000000	100.8123170

ICOSAHEDRON CLASS II COORDINATES: Frequencies 8, 4, 2

2ν	4ν	8ν	φ	θ
0,0	0,0	0,0	00.0000000	00.0000000
		1,0	00.0000000	8.4441473
		1,1	72.0000000	8.4441473
	1,0	2,0	00.0000000	17.7818206
		2,1	36.0000000	14.5454366
	1,1	2,2	72.0000000	17.7818206
		3,0	00.0000000	27.6098530
		3,1	22.3861776	23.5260254
		3,2	49.6138225	23.5260254
		3,3	72.0000000	27.6098530
1,0	2,0	4,0	00.0000000	37.3773681
		4,1	16.0353712	33.3269362
	2,1	4,2	36.0000000	31.7174744
		4,3	55.9646288	33.3269362
1,1	2,2	4,4	72.0000000	37.3773681
		5,0	00.0000000	47.4702086
		5,1	12.4463844	42.9758510
		5,2	27.7323014	40.7965529
		5,3	44.2676986	40.7965529
		5,4	59.5536157	42.9758510
	3,0	6,0	00.0000000	58.2825256
		6,1	10.9252950	53.3310985
	3,1	6,2	22.3861776	49.6967416
		6,3	36.0000000	48.8895123
	3,2	6,4	49.6138224	49.6967416

2ν	4ν	8ν	φ	θ
		7,0	00.0000000	69.0948426
		7,1	9.7323015	64.0653713
		7,2	19.9646288	59.8480232
		7,3	30.0743861	56.7273505
		7,4	41.9256139	56.7273505
2,0	4,0	8,0	00.0000000	79.1876831
		8,1	8.7723551	74.4432523
	4,1	8,2	18.0000000	70.0353712
		8,3	27.2276449	66.3105890
2,1	4,2	8,4	36.0000000	63.4349488
		9,0	00.0000000	88.9551982
		9,1	8.7723551	84.3401063
		9,2	18.0000000	79.7052785
		9,3	27.2276449	75.4545635
		9,4	36.0000000	71.8790961
	5,0	10.0	00.0000000	98.7832307
		10,1	8.7723551	94.5840965
	5,1	10,2	18.0000000	90.0000000
		10,3	27.2276449	85.4159036
	5,2	10,4	36.0000000	81.2167695
		11,0	00.0000000	108.1209040
		11,1	8.7723551	104.5454366
		11,2	18.0000000	100.2947216
		11,3	27.2276449	95.6598938
		11,4	36.0000000	91.0448019
3,0	6,0	12,0	00.0000000	116.5650512
		12,1	8.7723551	113.6894110
	6,1	12,2	18.0000000	109.9646288
		12,3	27.2276449	105.5567478
3,1	6,2	12,4	36.0000000	100.8123170

22. Tetrahedral Coordinates

Because its breakdowns entail an excessive number of different
component lengths, a problem compounded by the high frequencies required to obtain anything acceptably spherical, the
tetrahedron is of limited usefulness. Still, investigators may have a
use for data. Class II Method 1 is recommended for study. It has
orthogonal equators like the octahedron, permitting natural hemispheric truncation, and when so truncated provides large squarish
openings at ground level. (0,0 is not at the zenith.)

The designer might also consider omitting some of the shorter
struts that clutter the three-way vertices. So long as triangulation
is preserved this will do no structural damage and greatly improve
the symmetry.

Large families of intersecting equators — eight in the 4ν Class I
— permit numerous zenith locations and overhead symmetries.

Two Class I tables provide for the following frequencies: 16, 12,
8, 6, 4, 3. They cover one tetrahedron face, with corners at 0,0;
0, arccos −1/3; 120, arccos −1/3.

Two Class II Method 1 tables provide for sixteen, twelve, eight,
and six-frequency breakdowns. They cover the Class II triangle
with vertices at 0,0; 0, arccos 1/3; 120, arccos 1/3. See Diagram
22.1 for the location of this triangle; twelve of them make a
sphere.

22.1

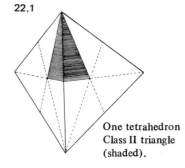

One tetrahedron
Class II triangle
(shaded).

TETRAHEDRON CLASS I COORDINATES: Frequencies 12, 6, 3

3ν	6ν	12ν	φ	θ
0,0	0,0	0,0	00.0000000	00.0000000
		1,0	00.0000000	5.0511525
		1,1	120.0000000	5.0511525
	1,0	2,0	00.0000000	11.4217537
		2,1	60.0000000	5.7681812
	1,1	2,2	120.0000000	11.4217537
		3,0	00.0000000	19.4712206
		3,1	30.0000000	11.5369590
		3,2	90.0000000	11.5369590
		3,3	120.0000000	19.4712206
1,0	2,0	4,0	00.0000000	29.4962085
		4,1	19.1066054	20.5141272
	2,1	4,2	60.0000000	15.7931691
		4,3	100.8933946	20.5141272
1,1	2,2	4,4	120.0000000	29.4962085
		5,0	00.0000000	41.4729343
		5,1	13.8978863	32.5125173
		5,2	40.8933947	25.0658292
		5,3	79.1066054	25.0658292
		5,4	106.1021138	32.5125173
		5,5	120.0000000	41.4729343
	3,0	6,0	00.0000000	54.7356103
		6,1	10.8933947	47.2058629
	3,1	6,2	30.0000000	39.2315205
		6,3	60.0000000	35.2643897
	3,2	6,4	90.0000000	39.2315205
		6,5	109.1066053	47.2058629
	3,3	6,6	120.0000000	54.7356103
		7,0	00.0000000	67.9982863
		7,1	8.9482756	63.0694073
		7,2	23.4132245	57.0210379
		7,3	46.1021138	51.8870735
		7,4	73.8978863	51.8870735
		7,5	96.5867756	57.0210380
		7,6	111.0517244	63.0694073
		7,7	120.0000000	67.9982863
2,0	4,0	8,0	00.0000000	79.9750121
		8,1	7.5890895	77.8296766
	4,1	8,2	19.1066054	75.0367825
		8,3	36.5867756	72.0247162
2,1	4,2	8,4	60.0000000	70.5287793
		8,5	83.4132244	72.0247162
	4,3	8,6	100.8933946	75.0367825
		8,7	112.4109105	77.8296766
2,2	4,4	8,8	120.0000000	79.9750121
		9,0	00.0000000	90.0000000
		9,1	6.587756	90.0000000
		9,2	16.1021138	90.0000000
		9,3	30.0000000	90.0000000
		9,4	49.1066054	90.0000000
		9,5	70.8933947	90.0000000
		9,6	90.0000000	90.0000000
		9,7	103.8978862	90.0000000
		9,8	113.4132244	90.0000000
		9,9	120.0000000	90.0000000
5,0	10,0		00.0000000	98.0494670
		10,1	5.8175256	99.3984525
5,1	10,2		13.8978863	101.0958033
		10,3	25.2849961	103.0884761
5,2	10,4		40.8933947	104.9632175
		10,5	60.0000000	105.7931691
5,3	10,6		79.1066054	104.9632175
		10,7	94.7150039	103.0884761
5,4	10,8		106.1021138	101.0958033
		10,9	114.1824744	99.3984525
5,5	10,10		120.0000000	98.0494670
		11,0	00.0000000	104.4200681
		11,1	5.2087191	106.5150758
		11,2	12.2163488	109.0624559
		11,3	21.7867893	112.0017137
		11,4	34.7150040	114.9379827
		11,5	51.0517245	116.9305927
		11,6	68.9482756	116.9305927
		11,7	85.2849961	114.9379827
		11,8	98.2132107	112.0017137
		11,9	107.7836512	109.0624558
		11,10	114.7912809	106.5150758
		11,11	120.0000000	104.4200681
3,0	6,0	12,0	00.0000000	109.4712206
		12,1	4.7150040	111.9343322
	6,1	12,2	10.8933947	114.8398925
		12,3	19.1066054	118.1255057
3,1	6,2	12,4	30.0000000	121.4821541
		12,5	43.8978863	124.1908638
	6,3	12,6	60.0000000	125.2643897
		12,7	76.1021138	124.1908638
3,2	6,4	12,8	90.0000000	121.4821541

(Continued)

3ν	6ν	12ν	φ	θ
		12,9	100.8933946	118.1255057
	6,5	12,10	109.1066053	114.8398925

3ν	6ν	12ν	φ	θ
		12,11	115.2849960	111.9343322
3,3	6,6	12,12	120.0000000	109.4712206

TETRAHEDRON CLASS I COORDINATES: Frequencies 16, 8, 4

4ν	8ν	16ν	φ	θ
0,0	0,0	0,0	00.0000000	00.0000000
		1,0	00.0000000	3.6780516
		1,1	120.0000000	3.6780516
	1,0	2,0	00.0000000	8.0494670
		2,1	60.0000000	4.0446912
	1,1	2,2	120.0000000	8.0494670
		3,0	00.0000000	13.2626760
		3,1	30.0000000	7.7493664
		3,2	90.0000000	7.7493664
		3,3	120.0000000	13.2626760
1,0	2,0	4,0	00.0000000	19.4712206
		4,1	19.1066054	13.1622883
	2,1	4,2	60.0000000	10.0249879
		4,3	100.8933946	13.1622883
1,1	2,2	4,4	120.0000000	19.4712206
		5,0	00.0000000	26.7972576
		5,1	13.8978863	20.0124169
		5,2	40.8933947	14.9632174
		5,3	79.1066054	14.9632174
		5,4	106.1021138	20.0124169
		5,5	120.0000000	26.7972576
	3,0	6,0	00.0000000	35.2643897
		6,1	10.8933947	28.3717844
	3,1	6,2	30.0000000	22.2076543
		6,3	60.0000000	19.4712206
	3,2	6,4	90.0000000	22.2076543
		6,5	109.1066053	28.3717844
	3,3	6,6	120.0000000	35.2643897
		7,0	00.0000000	44.7106224
		7,1	8.9482756	38.2169233
		7,2	23.4132245	31.6513993
		7,3	46.1021138	27.0171232
		7,4	73.8978863	27.0171232
		7,5	96.5687756	31.6513993
		7,6	111.0517244	38.2169233
		7,7	120.0000000	44.7106225

4ν	8ν	16ν	φ	θ
2,0	4,0	8,0	00.0000000	54.7356103
		8,1	7.5890895	49.2169021
	4,1	8,2	19.1066054	43.0887231
		8,3	36.5867756	37.6161120
2,1	4,2	8,4	60.0000000	35.2643897
		8,5	83.4132244	37.6161120
	4,3	8,6	100.8933946	43.0887232
		8,7	112.4109105	49.2169021
2,2	4,4	8,8	120.0000000	54.7356103
		9,0	00.0000000	64.7605982
		9,1	6.5867756	60.6661258
		9,2	16.1021138	55.8091362
		9,3	30.0000000	50.7684795
		9,4	49.1066054	47.2058629
		9,5	70.8933947	47.2058629
		9,6	90.0000000	50.7684795
		9,7	103.8978862	55.8091362
		9,8	113.4132244	60.6661258
		9,9	120.0000000	64.7605982
	5,0	10,0	00.0000000	74.2068310
		10,1	5.8175256	71.6832693
	5,1	10,2	13.8978863	68.5832360
		10,3	25.2849961	65.0620173
	5,2	10,4	40.8933947	61.8744943
		10,5	60.0000000	60.5037915
	5,3	10,6	79.1066054	61.8744943
		10,7	94.7150039	65.0620173
	5,4	10,8	106.1021138	68.5832360
		10,9	114.1824744	71.6832693
	5,5	10,10	120.0000000	74.2068310
		11,0	00.0000000	82.6739631
		11,1	5.2087191	81.5673284
		11,2	12.2163488	80.1975692
		11,3	21.7867893	78.5782463
		11,4	34.7150040	76.9115239
		11,5	51.0517245	75.7482449
		11,6	68.9482756	75.7482449
		11,7	85.2849961	76.9115239
		11,8	98.2132107	78.5782463
		11,9	107.7836512	80.1975692

(Continued)

4ν	8ν	16ν	φ	θ
		11,10	114.7912809	81.5673284
		11,11	120.0000000	82.6739631
3,0	6,0	12,0	00.0000000	90.0000000
		12,1	4.7150040	90.0000000
	6,1	12,2	10.8933947	90.0000000
		12,3	19.1066054	90.0000000
3,1	6,2	12,4	30.0000000	90.0000000
		12,5	43.8978863	90.0000000
	6,3	12,6	60.0000000	90.0000000
		12,7	76.1021138	90.0000000
3,2	6,4	12,8	90.0000000	90.0000000
		12,9	100.8933946	90.0000000
	6,5	12,10	109.1066054	90.0000000
		12,11	115.2849960	90.0000000
3,3	6,6	12,12	120.0000000	90.0000000
		13,0	00.0000000	96.2085447
		13,1	4.3066191	96.9911553
		13,2	9.8264298	97.9328899
		13,3	16.9960881	99.0406311
		13,4	26.3295035	100.2634279
		13,5	38.2132107	101.4217537
		13,6	52.4109105	102.1703234
		13,7	67.5890895	102.1703234
		13,8	81.7867893	101.4217537
		13,9	93.6704965	100.2634279
		13,10	103.0039119	99.0406311
		13,11	110.1735702	97.9328899
		13,12	115.6933809	96.9911553
		13;13	120.0000000	96.2085447
	7,0	14,0	00.0000000	101.4217537
		14,1	3.9632350	102.7203545
	7,1	14,2	8.9482756	104.2517550
		14,3	15.2953443	106.0231797
	7,2	14,4	23.4132245	107.9752839
		14,5	33.6704965	109.9074988
	7,3	14,6	46.1021138	111.4167141
		14,7	60.0000000	112.0017137
	7,4	14,8	73.8978863	111.4167141
		14,9	86.3295035	109.9074988
	7,5	14,10	96.5867756	107.9752839
		14,11	104.7046557	106.0231797
	7,6	14,12	111.0517244	104.2517550
		14,13	116.0367651	102.7203545
	7,7	14,14	120.0000000	101.4217537
		15,0	00.0000000	105.7931690
		15,1	3.6704965	107.4127302

4ν	8ν	16ν	φ	θ
		15,2	8.2132107	109.2863254
		15,3	13.8978863	111.4167140
		15,4	21.0517244	113.7466788
		15,5	30.0000000	116.1001388
		15,6	40.8933947	118.1255057
		15,7	53.4132245	119.3338742
		15,8	66.5867755	119.3338743
		15,9	79.1066054	118.1255057
		15,10	90.0000000	116.1001388
		15,11	98.9482756	113.7466788
		15,12	106.1021138	111.4167140
		15,13	111.7867893	109.2863254
		15,14	116.3295035	107.4127301
		15,15	120.0000000	105.7931691
4,0	8,0	16,0	00.0000000	109.4712206
		16,1	3.4179811	111.2776255
	8,1	16,2	7.5890895	113.3320386
		16,3	12.7305278	115.6320697
4,1	8,2	16,4	19.1066054	118.1255057
		16,5	26.9955084	120.6679177
	8,3	16,6	36.5867756	122.9789620
		16,7	47.7836512	124.6481469
4,2	8,4	16,8	60.0000000	125.2643897
		16,9	72.2163488	124.6481469
	8,5	16,10	83.4132244	122.9789620
		16,11	93.0044916	120.6679177
4,3	8,6	16,12	100.8933946	118.1255057
		16,13	107.2694722	115.6320697
	8,7	16,14	112.4109105	113.3320386
		16,15	116.5820189	111.2776255
4,4	8,8	16,16	120.0000000	109.4712206

TETRAHEDRON CLASS II COORDINATES: Frequencies 16, 8

8ν	16ν	φ	θ
0,0	0,0	00.0000000	00.0000000
	1,0	00.0000000	7.3260370
	1,1	120.0000000	7.3260370
1,0	2,0	00.0000000	15.7931690
	2,1	60.0000000	8.0494670
1,1	2,2	120.0000000	15.7931690
	3,0	00.0000000	25.2394018
	3,1	30.0000000	15.2251570
	3,2	90.0000000	15.2251570
	3,3	120.0000000	25.2394018
2,0	4,0	00.0000000	35.2643897
	4,1	19.1066054	25.0658292
2,1	4,2	60.0000000	19.4712206
	4,3	100.8933946	25.0658292
2,2	4,4	120.0000000	35.2643897
	5,0	00.0000000	45.2893775
	5,1	13.8978863	36.0707687
	5,2	40.8933947	28.1255057
	5,3	79.1066054	28.1255057
	5,4	106.1021138	36.0707687
	5,5	120.0000000	45.2893775

8ν	16ν	φ	θ
3,0	6,0	00.0000000	54.7356103
	6,1	10.8933947	47.2058629
3,1	6,2	30.0000000	39.2315205
	6,3	60.0000000	35.2643897
3,2	6,4	90.0000000	39.2315205
	6,5	109.1066053	47.2058629
3,3	6,6	120.0000000	54.7356103
	7,0	00.0000000	63.2027424
	7,1	8.9482756	57.5844268
	7,2	23.4132245	50.9542458
	7,3	46.1021138	45.5617593
	7,4	73.8978863	45.5617593
	7,5	96.5867756	50.9542458
	7,6	111.0517244	57.5844268
	7,7	120.0000000	63.2027424
4,0	8,0	00.0000000	70.5287793
	8,1	7.5890895	66.6679614
4,1	8,2	19.1066054	61.8744943
	8,3	36.5867756	57.0210380
4,2	8,4	60.0000000	54.7356103
	8,5	83.4132244	57.0210380
4,3	8,6	100.8933946	61.8744943
	8,7	112.4109105	66.6679614
4,4	8,8	120.0000000	70.5287793

TETRAHEDRON CLASS II COORDINATES: Frequencies 12, 6

6ν	12ν	φ	θ
0,0	0,0	00.0000000	00.0000000
	1,0	00.0000000	10.0249879
	1,1	120.0000000	10.0249879
1,0	2,0	00.0000000	22.0017137
	2,1	60.0000000	11.4217537
1,1	2,2	120.0000000	22.0017137
	3,0	00.0000000	35.2643897
	3,1	30.0000000	22.2076543
	3,2	90.0000000	22.2076543
	3,3	120.0000000	35.2643897
2,0	4,0	00.0000000	48.5270657
	4,1	19.1066054	36.8086651
2,1	4,2	60.0000000	29.4962085
	4,3	100.8933946	36.8086651
2,2	4,4	120.0000000	48.5270657

6ν	12ν	φ	θ
	5,0	00.0000000	60.5037915
	5,1	13.8978863	51.8870735
	5,2	40.8933947	43.0887231
	5,3	79.1066054	43.0887231
	5,4	106.1021138	51.8870735
	5,5	120.0000000	60.5037915
3,0	6,0	00.0000000	70.5287793
	6,1	10.8933947	65.1601075
3,1	6,2	30.0000000	58.5178459
	6,3	60.0000000	54.7356103
3,2	6,4	90.0000000	58.5178459
	6,5	109.1066053	65.1601076
3,3	6,6	120.0000000	70.5287793

23. Class II Method 3 Coordinates and Chord Factors

Chapter 10 has explained the advantage of this breakdown: a minimum set of different structural members (exactly n for n-frequency), obtained at some cost in symmetry. Since this advantage is confined to spherical contours, only spherical chord factors will be wanted. There is thus a unique useful set of chord factors for any frequency, and they are given here to seven places.

The following polyhedra and frequencies are provided for:

- Icosahedron: 16, 12, 8, 6, 4.
- Octahedron: 16, 12, 8, 6, 4.
- Tetrahedron: 16, 8.

The reader will remember that all Class II frequencies are even.

The tetrahedron is included as a curiosity: many of its dihedral angles are incised (greater than 180°). Situations are conceivable in which this fact — quite evident in 8ν — may have esthetic appeal.

The coordinate tables are included for two uses: obtaining truncation planes (see Chapter 14) and designing space frame structures (see Chapter 16).

The tables cover the same polyhedron zones as those for Class I Method 1. For details see the introductory pages to Chapters 19, 21, and 22.

ICOSAHEDRON CLASS II METHOD 3 COORDINATES: Frequencies 16, 8, 4

4ν	8ν	16ν	φ	θ
0,0	0,0	0,0	00.0000000	00.0000000
		1,0	00.0000000	4.8965007
		1,1	72.0000000	4.8965007
	1,0	2,0	00.0000000	9.7685316
		2,1	36.0000000	7.9293686
	1,1	2,2	72.0000000	9.7685316
		3,0	00.0000000	14.5928043
		3,1	22.3021252	12.2319064
		3,2	49.6978748	12.2319064
		3,3	72.0000000	14.5928043
1,0	2,0	4,0	00.0000000	19.3482660
		4,1	15.8580701	16.8349933
	2,1	4,2	36.0000000	15.8587372
		4,3	56.1419299	16.8349933
1,1	2,2	4,4	72.0000000	19.3482660
		5,0	00.0000000	24.0169264
		5,1	12.1757519	21.5075414
		5,2	27.5740342	20.0226254
		5,3	44.4259658	20.0226254
		5,4	59.8242481	21.5075414
		5,5	72.0000000	24.0169264
	3,0	6,0	00.0000000	28.5843958
		6,1	9.7926452	26.1653705
	3,1	6,2	22.0454269	24.4280040
		6,3	36.0000000	23.7881058
	3,2	6,4	49.9545731	24.4280040
		6,5	62.2073548	26.1653705
	3,3	6,6	72.000000	28.5843958
		7,0	00.0000000	33.0401240
		7,1	8.1159641	30.7655470
		7,2	18.1736779	28.9305071
		7,3	29.8428461	27.8896095
		7,4	42.1571539	27.8896095
		7,5	53.8263221	28.9305071
		7,6	63.8840359	30.7655470
		7,7	72.0000000	33.0401240
2,0	4,0	8,0	00.0000000	37.3773681
		8,1	6.8636824	35.2819581
	4,1	8,2	15.3112043	33.4502011
		8,3	25.2063518	32.1766730
2,1	4,2	8,4	36.0000000	31.7174744
		8,5	46.7936482	32.1766730
	4,3	8,6	56.6887957	33.4502011
		8,7	65.1363176	35.2819581
2,2	4,4	8,8	72.0000000	37.3773681
		9,0	00.0000000	41.9490488
		9,1	5.8853229	39.6978418
		9,2	13.1003088	37.9377757
		9,3	21.6019045	36.5527470
		9,4	31.0776846	35.7825182
		9,5	40.9223154	35.7825182
		9,6	50.3980955	36.5527470
		9,7	58.8996912	37.9377757
		9,8	66.1146771	39.6978418
	5,0	10,0	00.0000000	47.0530637
		10,1	5.4922539	44.5411604
	5,1	10,2	11.3309228	42.3608551
		10,3	18.7157055	40.9527399
	5,2	10,4	27.0726887	39.9903049
		10,5	36.0000000	39.6468430
	5,3	10,6	44.9273114	39.9903049
		10,7	53.2842945	40.9527399
	5,4	10,8	60.6690772	42.3608551
		11,0	00.0000000	52.5589890
		11,1	5.1317949	49.8595209
		11,2	10.5859950	47.4377847
		11,3	16.3430671	45.3310928
		11,4	23.7516178	44.2707079
		11,5	31.8182138	43.6894088
		11,6	40.1653899	43.6872157
		11,7	48.2483822	44.2707079
		11,8	55.6569329	45.3310928
3,0	6,0	12,0	00.0000000	58.2825256
		12,1	4.8068558	55.4941417
	6,1	12,2	9.9074472	52.9190485
		12,3	15.2945980	50.5988985
3,1	6,2	12,4	20.9459695	48.5708325
		12,5	28.2677630	47.8367118
	6,3	12,6	36.0000000	47.5762116
		12,7	43.7322370	47.8367118
3,2	6,4	12,8	51.0540305	48.5708325
		13,0	00.0000000	64.0060622
		13,1	4.5179372	61.2404806
		13,2	9.2989859	58.6204779
		13,3	14.3441422	56.1876974
		13,4	19.6435286	53.9823901
		13,5	25.1740844	52.0409579
		13,6	32.3293960	51.5981847

CLASS II METHOD 3 COORDINATES AND CHORD FACTORS   153

(Continued)

4ν	8ν	16ν	φ	θ
		13,7	39.6706040	51.5981847
		13,8	46.8259156	52.0409579
	7,0	14,0	00.0000000	69.5119876
		14,1	4.2640324	66.8763936
	7,1	14,2	8.7603869	64.3244498
		14,3	13.4953824	61.9046443
	7,2	14,4	18.4669970	59.6207795
		14,5	23.6648165	57.5452992
	7,3	14,6	29.0663431	55.7016224
		14,7	36.0000000	55.5055802
	7,4	14,8	42.9336569	55.7016224
		15,0	00.0000000	74.6160024
		15,1	4.0432955	72.1972961
		15,2	8.2890407	69.8132561
		15,3	12.7463540	67.4929229
		15,4	17.4192926	65.2685552
		15,5	22.3051890	63.1743833
		15,6	27.3930501	61.2449007
		15,7	32.6623267	59.5128088
		15,8	39.3376733	59.5128088
4,0	8,0	16,0	00.0000000	79.1876831
		16,1	3.8535036	77.0443133
	8,1	16,2	7.8810779	74.9024556
		16,3	12.0934305	72.7823469
4,1	8,2	16,4	16.4979737	70.7078972
		16,5	21.0975741	68.7062138
	8,3	16,6	25.8892192	66.8067019
		16,7	30.8627653	65.0397334
4,2	8,4	16,8	36.0000000	63.4343488
		17,0	00.0000000	83.5249272
		17,1	4.0021948	81.4976472
		17,2	8.1485481	79.4624774
		17,3	12.4430434	77.4387733
		17,4	16.8862381	75.4488400
		17,5	21.4745345	73.5172947
		17,6	26.1995453	71.6700912
		17,7	31.0476694	69.9332565
		17,8	36.0000000	68.3314495
	9,0	18,0	00.0000000	87.9806554
		18,1	4.1510787	86.0627193
	9,1	18,2	8.4138892	84.1237497
		18,3	12.7863749	82.1822215
	9,2	18,4	17.2634283	80.2590088
		18,5	21.8366543	78.3766542
	9,3	18,6	26.4942941	76.5583847

4ν	8ν	16ν	φ	θ
		18,7	31.2213584	74.8269602
	9,4	18,8	36.0000000	73.2034805
		19,0	00.0000000	92.5481249
		19,1	4.3015936	90.7289561
		19,2	8.6796460	88.8718118
		19,3	13.1267943	86.9945827
		19,4	17.6334681	85.1172037
		19,5	22.1880606	83.2608589
		19,6	26.7878296	81.4518342
		19,7	31.3863507	79.6963344
		19,8	36.0000000	78.0277532
5,0	10,0	20,0	00.0000000	97.2167853
		20,1	4.4552415	95.4819437
	10,1	20,2	8.9483793	93.6886019
		20,3	13.4675589	91.8546905
5,1	10,2	20,4	18.0000000	90.0000000
		20,5	22.5224410	88.1453096
	10,3	20,6	27.0516207	86.3113981
		20,7	31.5447585	84.5180564
5,2	10,4	20,8	36.0000000	82.7832148
		21,0	00.0000000	101.9722470
		21,1	4.6136493	100.3036656
		21,2	9.2227656	98.5529930
		21,3	13.8119394	96.7391412
		21,4	18.3665319	94.8827963
		21,5	22.8732057	93.0054174
		21,6	27.3203540	91.1281883
		21,7	31.6984064	89.2710441
		21,8	36.0000000	87.4518752
	11,0	22,0	00.0000000	106.7965196
		22,1	4.7786416	105.1730398
	11,1	22,2	9.5057059	103.4416154
		22,3	14.1724012	101.6307219
	11,2	22,4	18.7365717	99.7409913
		22,5	23.2136251	97.8177787
	11,3	22,6	27.5861108	95.8762505
		22,7	31.8489213	93.9372808
	11,4	22,8	36.0000000	92.0193447
		23,0	00.0000000	111.6685505
		23,1	4.9523306	110.0667436
		23,2	9.8004547	108.3299088
		23,3	14.5254655	106.4827054
		23,4	19.1137619	104.5511601
		23,5	23.5569566	102.5612268
		23,6	27.8514519	100.5375226

(Continued)

4ν	8ν	16ν	φ	θ
		23,7	31.9978052	98.5023528
		23,8	36.0000000	96.4750729
6,0	12,0	24,0	00.0000000	116.5650512
		24,1	5.1372347	114.9602668
	12,1	24,2	10.1107808	113.1932982

4ν	8ν	16ν	φ	θ
		24,3	14.9024259	111.2937863
6,1	12,2	24,4	19.5020264	109.2921029
		24,5	23.9065696	107.2176532
	12,3	24,6	28.1189221	105.0975445
		24,7	32.1464964	102.9556867
6,2	12,4	24,8	36.0000000	100.8123170

ICOSAHEDRON CLASS II METHOD 3 COORDINATES: Frequencies 12, 6

6ν	12ν	φ	θ
0,0	0,0	00.0000000	00.0000000
	1,0	00.0000000	6.5244151
	1,1	72.0000000	6.5244151
1,0	2,0	00.0000000	12.9913757
	2,1	36.0000000	10.5724915
1,1	2,2	72.0000000	12.9913757
	3,0	00.0000000	19.3482660
	3,1	22.2362326	16.3030135
	3,2	49.7637674	16.3030135
	3,3	72.0000000	19.3482660
2,0	4,0	00.0000000	25.5512604
	4,1	15.7184691	22.4080487
2,1	4,2	36.0000000	21.1449829
	4,3	56.2815310	22.4080487
2,2	4,4	72.0000000	25.5512604
	5,0	00.0000000	31.5677896
	5,1	11.9615443	28.5589568
	5,2	27.4479614	26.6874397
	5,3	44.5520386	26.6874397
	5,4	60.0384557	28.5589568
	5,5	72.0000000	31.5677896
3,0	6,0	00.0000000	37.3773681
	6,1	9.5030365	34.6279175
3,1	6,2	21.7718193	32.5212749
	6,3	36.0000000	31.7174744
3,2	6,4	50.2281807	32.5212749
	6,5	62.4969634	34.6279175
3,3	6,6	72.0000000	37.3773681
	7,0	00.0000000	43.5964648
	7,1	7.7495361	40.5491327
	7,2	17.7436617	38.4355945

6ν	12ν	φ	θ
	7,3	29.6538469	37.1731351
	7,4	42.3461531	37.1731351
	7,5	54.2563383	38.4355945
	7,6	64.2504640	40.5491327
4,0	8,0	00.0000000	50.6894439
	8,1	7.0718179	47.2244975
4,1	8,2	14.7192966	44.3091697
	8,3	24.7941682	42.8394006
4,2	8,4	36.0000000	42.2899659
	8,5	47.2058318	42.8394006
4,3	8,6	57.2807034	44.3091697
	9,0	00.0000000	58.2825256
	9,1	6.4744904	54.6099560
	9,2	13.4680783	51.3415419
	9,3	20.9459694	48.5708325
	9,4	30.8186086	47.6930172
	9,5	41.1813914	47.6930172
	9,6	51.0540306	48.5708325
5,0	10,0	00.0000000	65.8756073
	10,1	5.9618466	62.2565770
5,1	10,2	12.3756033	58.8817485
	10,3	19.2370783	55.8469789
5,2	10,4	26.5068526	53.2419865
	10,5	36.0000000	52.8624574
5,3	10,6	45.4931474	53.2419865
	11,0	00.0000000	72.9685863
	11,1	5.5310891	69.6460760
	11,2	11.4448720	66.4298070
	11,3	17.7542810	63.3987595
	11,4	24.4501339	60.6374660
	11,5	31.4941917	58.2281480
	11,6	40.5058083	58.2281480

(Continued)

6ν	12ν	φ	θ
6,0	12,0	00.0000000	79.1876830
	12,1	5.1760132	76.3292716
6,1	12,2	10.6682436	73.4852294
	12,3	16.4979736	70.7078973
6,2	12,4	22.6737941	68.0601958
	12,5	29.1856566	65.6121306
6,3	12,6	36.0000000	63.4349488
	13,0	00.0000000	84.9972617
	13,1	5.4318507	82.3405545
	13,2	11.1026671	79.6738306
	13,3	17.0129974	77.0455288
	13,4	23.1516172	74.5113243
	13,5	29.4936621	72.1302270
	13,6	36.0000000	69.9593641
7,0	14,0	00.0000000	91.0137909
	14,1	5.6881081	88.5382921
7,1	14,2	11.5304941	86.0165470
	14,3	17.5106870	83.4938802
7,2	14,4	23.6043027	81.0207415
	14,5	29.7797810	78.6487727
7,3	14,6	36.0000000	76.4263245
	15,0	00.0000000	97.2167853
	15,1	5.9493053	94.8898244
	15,2	11.9590467	92.4693098
	15,3	18.0000000	90.0000000

6ν	12ν	φ	θ
	15,4	24.0409533	87.5306904
	15,5	30.0506948	85.1101757
	15,6	36.0000000	82.7832149
8,0	16,0	00.0000000	103.5736756
	16,1	6.2202190	101.3512274
8,1	16,2	12.3956973	98.9792587
	16,3	18.4893130	96.5061200
8,2	16,4	24.4695059	93.9834532
	16,5	30.3118919	91.4617081
8,3	16,6	36.0000000	88.9862092
	17,0	00.0000000	110.0406361
	17,1	6.5063380	107.8697730
	17,2	12.8483828	105.4886757
	17,3	18.9870026	102.9544713
	17,4	24.8973329	100.3261696
	17,5	30.5681493	97.6594457
	17,6	36.0000000	95.0027385
9,0	18,0	00.0000000	116.5650512
	18,1	6.8143434	114.3878695
9,1	18,2	13.3262058	111.9398043
	18,3	19.5020264	109.2921029
9,2	18,4	25.3317565	106.5147707
	18,5	30.8238967	103.6707285
9,3	18,6	36.0000000	100.8123170

OCTAHEDRON CLASS II METHOD 3 COODINATES: Frequencies 16, 8, 4

4ν	8ν	16ν	φ	θ
0,0	0,0	0,0	00.0000000	00.0000000
		1,0	00.0000000	7.9295893
		1,1	90.0000000	7.9295893
	1,0	2,0	00.0000000	15.7115351
		2,1	45.0000000	11.2500000
	1,1	2,2	90.0000000	15.7115351
		3,0	00.0000000	23.2191868
		3,1	26.3422998	17.7536531
		3,2	63.6577002	17.7536531
		3,3	90.0000000	23.2191868
1,0	2,0	4,0	00.0000000	30.3611934
		4,1	17.9877140	24.9352975
	2,1	4,2	45.0000000	22.5000000
		4,3	72.0122860	24.9352975
1,1	2,2	4,4	90.0000000	30.3611934
		5,0	00.0000000	37.0861048
		5,1	13.3753647	32.1176318
		5,2	33.2539055	28.6320487
		5,3	56.7460945	28.6320487
		5,4	76.6246353	32.1176318
		5,5	90.0000000	37.0861048
	3,0	6,0	00.0000000	43.3786471
		6,1	10.4404638	39.0542204
	3,1	6,2	25.6511069	35.3052683
		6,3	45.0000000	33.7500000
	3,2	6,4	64.3488931	35.3052683
		6,5	79.5595362	39.0542204
	3,3	6,6	90.0000000	43.3786471
		7,0	00.0000000	49.2514146
		7,1	8.3851765	45.6358336
		7,2	20.4121899	42.0667135
		7,3	36.2169469	39.7068033
		7,4	53.7830531	39.7068033
		7,5	69.5878101	42.0667135
		7,6	81.6148235	45.6358336
		7,7	90.0000000	49.2514146
2,0	4,0	8,0	00.0000000	54.7356103
		8,1	6.8434581	51.8211192
	4,1	8,2	16.5779208	48.6694954
		8,3	29.5758522	46.0506036
2,1	4,2	8,4	45.0000000	45.0000000
		8,5	60.4241478	46.0506036

4ν	8ν	16ν	φ	θ
	4,3	8,6	73.4220792	48.6694954
		8,7	83.1565419	51.8211192
2,2	4,4	8,8	90.0000000	54.7356103
		9,0	00.0000000	60.9447533
		9,1	5.6250000	57.6092324
		9,2	13.6243528	54.9821198
		9,3	24.4175923	52.4651854
		9,4	37.7735032	50.8489027
		9,5	52.2264968	50.8489027
		9,6	65.5824077	52.4651854
		9,7	76.3756472	54.9821198
		9,8	84.3750000	57.6092324
		9,9	90.0000000	60.9447533
	5,0	10,0	00.0000000	69.0589795
		10,1	5.62500000	64.7609631
	5,1	10,2	11.2500000	60.9447532
		10,3	20.2857745	58.7432366
	5,2	10,4	31.7953702	56.9555030
		10,5	45.0000000	56.2500000
	5,3	10,6	58.2046298	56.9555030
		10,7	69.7142255	58.7432366
	5,4	10,8	78.7500000	60.9447532
		10,9	84.3750000	64.7609631
	5,5	10,10	90.0000000	69.0589795
		11,0	00.0000000	78.9608056
		11,1	5.6250000	73.8127966
		11,2	11.2500000	69.0589794
		11,3	16.8750000	64.7609631
		11,4	26.7810227	63.0823973
		11,5	38.6581468	62.0209628
		11,6	51.3418532	62.0209628
		11,7	63.2189773	63.0823973
		11,8	73.1250000	64.7609631
		11,9	78.7500000	69.0589794
		11,10	84.3750000	73.8127966
		11,11	90.0000000	78.9608056
3,0	6,0	12,0	00.0000000	90.0000000
		12,1	5.6250000	84.4019135
	6,1	12,2	11.2500000	78.9608056
		12,3	16.8750000	73.8127966
3,1	6,2	12,4	22.5000000	69.0589795
		12,5	33.0763472	67.9384490
	6,3	12,6	45.0000000	67.5000000
		12,7	56.9236528	67.9384490
3,2	6,4	12,8	67.5000000	69.0589795
		12,9	73.1250000	73.8127966

(Continued)

4ν	8ν	16ν	φ	θ
	6,5	12,10	78.7500000	78.9608056
		12,11	84.3750000	84.4019135
3,3	6,6	12,12	90.0000000	90.0000000
		13,1	5.6250000	95.5980865
		13,2	11.2500000	90.0000000
		13,3	16.8750000	84.4019135
		13,4	22.5000000	78.9608056
		13,5	28.1250000	73.8127966
		13,6	39.1517239	73.2078753
		13,7	50.8482761	73.2078753
		13,8	61.8750000	73.8127966
		13,9	67.5000000	78.9608056
		13,10	73.1250000	84.4019135
		13,11	78.7500000	90.0000000
		13,12	84.3750000	95.5980865
		14,3	16.8750000	95.5980865
	7,2	14,4	22.5000000	90.0000000

4ν	8ν	16ν	φ	θ
		14,5	28.1250000	84.4019135
	7,3	14,6	33.7500000	78.9608055
		14,7	45.0000000	78.7500000
	7,4	14,8	56.2500000	78.9608055
		14,9	61.8750000	84.4019135
	7,5	14,10	67.5000000	90.0000000
		14,11	73.1250000	95.5980865
		15,5	28.1250000	95.5980865
		15,6	33.7500000	90.0000000
		15,7	39.3750000	84.4019135
		15,8	50.6250000	84.4019135
		15,9	56.2500000	90.0000000
		15,10	61.8750000	95.5980865
		16,7	39.3750000	95.5980865
4,2	8,4	16,8	45.0000000	90.0000000
		16,9	50.6250000	95.5980865

OCTAHEDRON CLASS II METHOD 3 COORDINATES: Frequencies 12, 6

6ν	12ν	φ	θ
0,0	0,0	00.0000000	00.0000000
	1,0	00.0000000	10.5468389
	1,1	90.0000000	10.5468389
1,0	2,0	00.0000000	20.7535710
	2,1	45.0000000	15.0000000
1,1	2,2	90.0000000	20.7535710
	3,0	00.0000000	30 3611934
	3,1	26.1664470	23.6363461
	3,2	63.8335531	23.6363461
	3,3	90.0000000	30.3611934
2,0	4,0	00.0000000	39.2315205
	4,1	17.6321948	33.0285060
2,1	4,2	45.0000000	30.0000000
	4,3	72.3678052	33.0285060
2,2	4,4	90.0000000	39.2315205
	5,0	00.0000000	47.3388280
	5,1	12.8454522	42.1874744
	5,2	32.8982858	38.1236489
	5,3	57.1017142	38.1236489
	5,4	77.1545478	42.1874744
	5,5	90.0000000	47.3388280

6ν	12ν	φ	θ
3,0	6,0	00.0000000	54.7356103
	6,1	9.7356103	50.7684795
3,1	6,2	24.8960906	46.7997474
	6,3	45.0000000	45.0000000
3,2	6,4	65.1039094	46.7997474
	6,5	80.2643897	50.7684795
3,3	6,6	90.0000000	54.7356103
	7,0	00.0000000	63.4349488
	7,1	7.5000000	58.6685572
	7,2	19.2491090	55.3506843
	7,3	35.6571298	52.8688870
	7,4	54.3428702	52.8688870
	7,5	70.7508910	55.3506843
	7,6	82.5000000	58.6685572
	7,7	90.0000000	63.4349488
4,0	8,0	00.0000000	75.4891813
	8,1	7.5000000	69.0589796
4,1	8,2	15.0000000	63.4349488
	8,3	28.3607780	61.0497389
4,2	8,4	45.0000000	60.0000000
	8,5	61.6392220	61.0497389
4,3	8,6	75.0000000	63.4349488
	8,7	82.5000000	69.0589796
4,4	8,8	90.0000000	75.4891813

(Continued)

6ν	12ν	φ	θ
	9,0	00.0000000	90.0000000
	9,1	7.5000000	82.5634423
	9,2	15.0000000	75.4891814
	9,3	22.5000000	69.0589795
	9,4	36.9584436	67.6994773
	9,5	53.0415564	67.6994773
	9,6	67.5000000	69.0589795
	9,7	75.0000000	75.4891814
	9,8	82.5000000	82.5634423
	9,9	90.0000000	90.0000000
	10,1	7.5000000	97.4365577
5,1	10,2	15.0000000	90.0000000
	10,3	22.5000000	82.5634423
5,2	10,4	30.0000000	75.4891813

6ν	12ν	φ	θ
	10,5	45.0000000	75.0000000
5,3	10,6	60.0000000	75.4891813
	10,7	67.5000000	82.5634423
5,4	10,8	75.0000000	90.0000000
	10,9	82.5000000	97.4365577
	11,3	22.5000000	97.4365577
	11,4	30.0000000	90.0000000
	11,5	37.5000000	82.5634423
	11,6	52.5000000	82.5634423
	11,7	60.0000000	90.0000000
	11,8	67.5000000	97.4365577
	12,5	37.5000000	97.4365576
6,3	12,6	45.0000000	90.0000000
	12,7	52.5000000	97.4365576

TETRAHEDRON CLASS II METHOD 3 COORDINATES: Frequencies 16, 8

8ν	16ν	φ	θ
0,0	0,0	00.0000000	00.0000000
	1,0	00.0000000	13.4941562
	1,1	120.0000000	13.4941562
1,0	2,0	00.0000000	25.9638679
	2,1	60.0000000	13.6839026
1,1	2,2	120.0000000	25.9638679
	3,0	00.0000000	36.8258015
	3,1	29.5237783	23.4810159
	3,2	90.4762218	23.4810159
	3,3	120.0000000	36.8258015
2,0	4,0	00.0000000	45.9929728
	4,1	18.2885796	34.7380158
2,1	4,2	60.0000000	27.3678052
	4,3	101.7114204	34.7380158
2,2	4,4	120.0000000	45.9929728
	5,0	00.0000000	53.6670710
	5,1	12.7923764	45.0212703
	5,2	39.8464486	35.9116199
	5,3	80.1535514	35.9116199
	5,4	107.2076236	45.0212703
	5,5	120.0000000	53.6670710

8ν	16ν	φ	θ
3,0	6,0	00.0000000	60.1382758
	6,1	9.5163820	53.8464444
3,1	6,2	28.0397784	45.7483307
	6,3	60.0000000	41.0517077
3,2	6,4	91.9602216	45.7483307
	6,5	110.4836180	53.8464444
3,3	6,6	120.0000000	60.1382758
	7,0	00.0000000	65.6824939
	7,1	7.3023885	61.2902881
	7,2	20.6967847	55.0333000
	7,3	44.4574567	48.9532232
	7,4	75.5425433	48.9532232
	7,5	99.3032153	55.0333000
	7,6	112.6976114	61.1902881
	7,7	120.0000000	65.6824939
4,0	8,0	00.0000000	70.5287793
	8,1	5.6705920	67.5921201
4,1	8,2	15.7201505	63.1494221
	8,3	33.3511053	57.7074367
4,2	8,4	60.0000000	54.7356103
	8,5	86.6488947	57.7074367
4,3	8,6	104.2798495	63.1494221
	8,7	114.3294080	67.5921201
4,4	8,8	120.0000000	70.5287793

CHORD FACTORS: CLASS II METHOD 3

ICOSAHEDRON

4ν

0,0.1,0	.3360887
1,0/1,1	.3894775
1,0/2,0	.3133706
2,0/2,1	.3628433

6ν

0,0/1,0	.2262569
1,0/1,1	.2642734
1,0/2,0	.2187727
2,0/2,1	.2556190
2,0/3,0	.2060383
3,0/3,1	.2318173

8ν

0,0/1,0	.1702866
1,0/1,1	.1994570
1,0/2,0	.1670032
2,0/2,1	.1956775
2,0/3,0	.1610264
3,0/3,1	.1849139
3,0/4,0	.1533158
4,0/4,1	.1686722

12ν

0,0/1,0	.1138110
1,0/1,1	.1335761
1,0/2,0	.1128099
2,0/2,1	.1324273
2,0/3,0	.1108918
3,0/3,1	.1290636
3,0/4,0	.1082098
4,0/4,1	.1237168
4,0/5,0	.1049600
5,0/5,1	.1167275
5,0/6,0	.1013528
6,0/6,1	.1084904

16ν

0,0/1,0	.0854341
1,0/1,1	.1003421
1,0/2,0	.0850074
2,0/2,1	.0998530
2,0/3,0	.0841746
3,0/3,1	.0984061
3,0/4,0	.0829746
4,0/4,1	.0960596
4,0/5,0	.0814610
5,0/5,1	.0929039
5,0/6,0	.0796963
6,0/6,1	.0890524
6,0/7,0	.0777475
7,0/7,1	.0846314
7,0/8,0	.0756811
8,0/8,1	.0797697

OCTAHEDRON

4ν

0,0/1,0	.5237247
1,0/1,1	.7148135
1,0/2,0	.4222132
2,0/2,1	.6058109

6ν

0,0/1,0	.3602412
1,0/1,1	.5011256
1,0/2,0	.3211053
2,0/2,1	.4595058
2,0/3,0	.2697726
3,0/3,1	.3602412

8ν

0,0/1,0	.2733597
1,0/1,1	.3829609
1,0/2,0	.2549889
2,0/2,1	.3634589
2,0/3,0	.2267091
3,0/3,1	.3131456
3,0/4,0	.1978921
4,0/4,1	.2493395

12ν

0,0/1,0	.1838173
1,0/1,1	.2588566
1,0/2,0	.1779056
2,0/2,1	.2525852
2,0/3,0	.1674883
3,0/3,1	.2351521
3,0/4,0	.1546619
4,0/4,1	.2099983
4,0/5,0	.1413812
5,0/5,1	.1810987
5,0/6,0	.1290086
6,0/6,1	.1516863

16ν

0,0/1,0	.1382870
1,0/1,1	.1950993
1,0/2,0	.1357162
2,0/2,1	.1923724
2,0/3,0	.1309395
3,0/3,1	.1845520
3,0/4,0	.1245708
4,0/4,1	.1726045
4,0/5,0	.1173045
5,0/5,1	.1578201
5,0/6,0	.1097704
6,0/6,1	.1415017
6,0/7,0	.1024543
7,0/7,1	.1247402
7,0/8,0	.0956807
8,0/8,1	.1083170

TETRAHEDRON

8ν

0,0/1,0	.4492876
1,0/1,1	.7582992
1,0/2,0	.3477966
2,0/2,1	.6598375
2,0/3,0	.2462556
3,0/3,1	.4566979
3,0/4,0	.1811001
4,0/4,1	.2819394

16ν

0,0/1,0	.2349735
1,0/1,1	.4041675
1,0/2,0	.2172083
2,0/2,1	.3864314
2,0/3,0	.1892927
3,0/3,1	.3406228
3,0/4,0	.1598267
4,0/4,1	.2826089
4,0/5,0	.1338382
5,0/5,1	.2258662
5,0/6,0	.1128838
6,0/6,1	.1769733
6,0/7,0	.0967271
7,0/7,1	.1372546
7,0/8,0	.0845584
8,0/8,1	.1056273

CLASS II METHOD 3 COORDINATES AND CHORD FACTORS   161

Writing Class II Method 3 Coordinates

I haven't been able to find an easier method than the following. Diagram I.1 shows a 6ν example.

1. Divide the vertical median of a Class II triangle face into $n/2$ equal parts for n-frequency. This median is:

octahedron: $45°$,
icosahedron: arc tan $(\tau - 1)/2$, or $31.7174744°$,
tetrahedron: arc tan $\sqrt{2}$, or $54.7356103°$.

For the 6ν icosa shown, the 3 divisions are bounded at 10.5724915 and 21.1449829. The first of these isn't a vertex; the second gives us the θ of 2,1. (Every second division point will fall on a vertex.)

2. Great-circle arcs from these division points meet the triangle's left and right sides at $90°$. To find the θ of the intersection we use the equation

$$\tan \theta_1 = \tan \theta / \cos A. \qquad [\text{Eq. I.1}]$$

Here A is $1/2$ the angle at a polyhedron vertex (that is, $45°$ for the octa, $36°$ for the icosa, $60°$ for the tetra). So in the example, the θ of 1,0 and 1,1 is

arc tan $(\tan 10.5724915 / \cos 36) = 12.9913757$.

Similarly, the θ of 2,0 and 2,2 is

arc tan $(\tan 21.1449829 / \cos 36) = 25.5512604$.

When we have completed this process we know the θ of every point down the right and left edges and along the vertical median. And for the icosa the corresponding ϕ values are of course 72, 0, and 36, respectively.

3. All other points are obtained by rotating the Class II triangle

A1.1

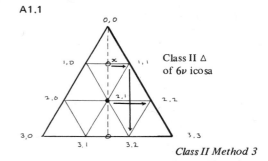

Class II △
of 6ν icosa

Class II Method 3

1. Points x and 2,1 obtained by trisecting median.
2. Points 1,0 1,1 and 2,0 2,1 obtained from x and 2,1 by Equation 1.1.
3. Points 3,1 3,2 obtained by rotating the system downward.

downward the length of one diamond. Thus in the example, if the diamond 0,0-1,0-2,1-1,1 is slid downward its own length till 0,0 falls on the spot vacated by 2,1, then 1,0 will move to 3,1 and 1,1 to 3,2. To accomplish this we use the rotation formulae, Equation 14.1 and Equation 14.2, and the point we move to 0,0 is the one with the same θ as 2,1, but lying on the ϕ meridian 180° away (that is, point 216, 21.1449829). When we do this we find that 1,0 moves to 21.7718193, 32.5212749, and 1,1 moves to 50.2281807, 32.5212749. These are 3,1 and 3,2, respectively, which accounts for all the points in the 6ν Class II triangle.

Diagram I.2 shows a detail to watch. In this 10ν example, we move the diamond 0,0-1,0-2,1-1,1, as before, and get 3,1 from 1,0, then 3,2 from 1,1, also 4,1 and 4,3 from 2,0 and 2,2, finally 5,1 and 5,4 from 3,0 and 3,3. What about 5,2 and 5,3? We could get them in turn from 3,1 and 3,2, but these points themselves have already been obtained by rotation, and we may have reason to be nervous about accumulating errors, if not here, certainly in some very-high-frequency example in which third- and fourth-generation rotations are conceivable. A better strategy is to move the diamond 0,0-2,0-4,2-2,2, whereupon 1,0 will give us 5,2 and 1,1 will give us 5,3. By suitable choice of a diamond, and faith in symmetry, it is always possible to avoid deriving a point from a point that was itself derived by rotation.

Finally, the reader should notice a quirk in the rotation formulae, of consequence only when, as here, we are relocating points within a system but not moving the whole system. In effect, the two equations answer the question, where does point p,q go when another point m,n is moved to 0,0? A little reflection will convince us that this question has an infinite number of answers, since it specifies only that the distance between two points shall not change. When m,n is at 0,0 then the θ of p,q will simply be the distance that separates them; in fact Equation 14.1 is nothing but the distance formula (Equation 9.2) with the second-power terms removed. But the ϕ of p,q can have any value whatever: p,q can lie anywhere along the lesser circle of center 0,0 and radius θ. It is only when we add a third point, r,s, that the ϕ values become

A1.2

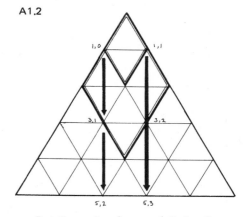

Rotating system downward the length of the *longer* diamond will obtain 5,2 and 5,3 directly from 1,0 and 1,1, instead of from 3,1 and 3,2, which were themselves obtained by rotation.

meaingful, and then all they specify is that the relationship between the three points shall not have changed; they say nothing
about the icosahedron, or whatever it was, from which we took
the three points. If all three points have swung, say, 36° westward,
their private interrelations are still unvaried — which is what the
rotation formulae set out to achieve — though they have all three
come unmoored from the icosa face where we found them.

And in fact Equation 14.2 has a habit of swinging things about
in this way, seeking as it does a convenient axis of rotation without consulting our preference. But since it rotates everything by
the same amount, we have only to find out what it is doing and we
shall know what correction to apply. In the 6ν example above,
sliding the diamond 0,0-1,0-2,1-1,1 straight down the Class II
triangle's median ought to bring 0,0 to 2,1 — that is, to 36,
21.1449829. In fact 0,0 ends up at 0, 21.1449829, which means
that the system swings 36° westward. So we apply a correction of
+36° to all ϕ values yielded by Equation 14.2.

This fact did not trouble us when we were using the rotation
formulae to derive new chord factors for truncation planes, since
we always moved both of the points involved and never referred
from them to points of the system that had not undergone
rotation.

Calculator Routines

Before the introduction of the "scientific" pocket calculator in 1972, spherical coordinates were quite impractical. The basic distance formula (for chord factors) required looking up six different trig functions, manipulating these unwieldy numbers, and taking a square root. Even if a slide rule gave sufficient accuracy (which it doesn't, except for crude geodesic structures) the need to add or subtract would drive its user to pencil and paper at least three times in the course of computing one chord factor. The calculator gives multiplace accuracy in less than a minute.

Routines for handling our principal equations on the Hewlett-Packard HP-35 are given below. This is a comparatively simple machine with a 4-level stack but only one storage register, and one must sometimes manipulate the stack a good deal to avoid reentering data. The reader may find better procedures of his own. The routines can also be used on other Hewlett-Packard machines such as the HP-21 and HP-45, with the following modifications. (1) Remember to insert shift prefixes where appropriate. (2) Use the alternate routines provided for Equations 13.3, 13.4 and 13.7, to accommodate a power function of the form y^x instead of the HP-35's x^y. (3) On the HP-45 with its addressable storage registers, remember to supply an address for each STO. You may want to rewrite the routines using more than one register and avoiding all data reentry.

The HP-35 routines are read left to right. Each block is either the name of a piece of data or the name of a machine key (↑ means "enter"). Thus the first seven items in the Equation 9.2 (chord factors of spheres) routine tell you to key the value of θ_1, press "enter," press "cos," key the value of θ_2, press "STO," press "cos," press "X."

Chord Factors of Spheres (Equation 9.2)

$\theta_1 \uparrow \cos \theta_2$ STO $\cos \times x \leftrightarrows y \sin$ RCL $\sin \times \phi_1 \uparrow \phi_2 - \cos \times + 2 \times$
CHS $\uparrow 2 + \sqrt{\ }$

Chord Factors of Nonspherical Structures (Equation 9.1)

$\theta_1 \uparrow \cos \theta_2$ STO $\cos \times x \leftrightarrows y \sin$ RCL $\sin \times \phi_1 \uparrow \phi_2 - \cos \times + 2 \times$
CHS STO $r_1 \uparrow \uparrow$ RCL $\times$ STO R$\downarrow \uparrow \times r_2 \uparrow \uparrow$ RCL $\times$ STO R$\downarrow \uparrow \times +$ RCL
$+ \sqrt{\ }$

Radius of an Ellipse (Equation 13.1)

E $\uparrow \times \uparrow \uparrow \theta$ STO $\sin \uparrow \times \times$ RCL $\cos \uparrow \times + \div \sqrt{\ }$

Radius of a Superellipse (Equation 13.3)

$2.5 \uparrow$ E $x^y \uparrow \uparrow \uparrow 2.5 \uparrow \theta \cos$ STO arc $\cos \sin x^y \times 2.5$ RCL $x^y + \div$
$2.5 \ ^1/x \ x \leftrightarrows y \ x^y$

Alternative form, for y^x machines:

E $\uparrow 2.5 \ y^x \uparrow \uparrow \theta$ STO $\sin 2.5 \ y^x \times$ RCL $\cos 2.5 \ y^x + \div 2.5 \ ^1/x \ y^x$

Radius of a Supercircle (Equation 13.4)

$2.5 \uparrow \uparrow \uparrow \theta$ STO $\sin x^y$ RCL $\cos$ R$\downarrow \ x \leftrightarrows y$ R$\downarrow$ R$\downarrow$ R$\downarrow \ x^y + \ ^1/x \ x \leftrightarrows y \ x^y$

Alternative form, for y^x machines:

θ STO $\sin 2.5 \ y^x$ RCL $\cos 2.5 \ y^x + \ ^1/x \ 2.5 \ ^1/x \ y^x$

Radius When Two System Radii Are Unequal (Equation 13.7)

$n \uparrow \uparrow \uparrow \phi$ STO $\cos x^y$ RCL $\sin$ $E_1 \times$ R$\downarrow$ R$\downarrow$ R$\downarrow \ x \leftrightarrows y \ x^y + \theta$ STO $\sin$
R$\downarrow$ R$\downarrow$ R$\downarrow \ x \leftrightarrows y \ E_2 \times x^y \times n$ RCL $\cos E_1 \times x^y + \ ^1/x$ STO $n \uparrow \uparrow E_1$
$\uparrow E_2 \times x^y$ RCL $\times x \leftrightarrows y$
$^1/x \ x \leftrightarrows y \ x^y$

Alternative form, for y^x machines:

ϕ STO $\cos n \ y^x$ RCL $\sin E_1 \times n \ y^x + \theta$ STO $\sin E_2 \times n \ y^x \times$ RCL
$\cos E_1$ STO $\times n \ y^x + E_2$ RCL $\times n \ y^x \ x \leftrightarrows y \div n \ ^1/x \ y^x$

Rotation Formulae (Equation 14.1, Equation 14.2)

$\theta \uparrow \cos \theta_1$ STO $\cos \times x \leftrightarrows y \sin$ RCL $\sin \times \phi \uparrow \phi_1 - \cos \times +$ arc $\cos$
STO

(Write down value of θ_2 and continue:)

θ_1 sin ϕ ↑ ϕ_1 – sin × RCL sin ÷ arc sin

(Value of ϕ_2 is now on display.)

Face Angles (Equation 17.4, Equation 17.5, Equation 17.6)

a ↑ ↑ CHS × b STO ↑ × + c ↑ ↑ × R↓ R↓ R↓ + $x \leftrightarrows y$ ÷ 2 ÷ RCL ÷ arc
cos STO

(Write down value of Face Angle A and continue:)

sin b × a ÷ arc sin

(Write down value of Face Angle B and continue:)

RCL + 180 $x \leftrightarrows y$ –

(Value of Face Angle C is now on display.)

Dihedral Angles of Spheres (Equation 17.7)

B cos c × STO ↑ × RCL a × STO – 1 + $\sqrt{}$ 2 × c × B sin × $^1/x$ c ↑ ×
RCL – × arc cos

HP-65 Programs

Many of the programs that generated data for this book have been filed with the HP-65 Users' Library. *GEODESICS I: Icosa Coordinates* (02487A) produces Class I and Class II breakdown data for the #1 triangle and the Class II triangle respectively; the only input needed is the frequency. *GEODESICS VII: Icosa Class I Second Triangle* (02692) accepts the #1 triangle values as inputs and applies rotation formulae to obtain Triangle #2. *GEODESICS VIII: Class II Icosa Triangles 2, 3, 4, 5* (02701A) accepts data for the Class II triangle (#1 at the top of Diagram 12.5) and generates the coordinates of the other four. *GEODESICS VI: Octahedron Coordinates* (02691) gives Class I and Class II breakdown values for any frequency. *GEODESICS IX: Triacon Coordinates* (02734A) applies the methods outlined in Appendix 1 to the icosahedron, octahedron or tetrahedron. A reader who needs data for higher frequencies than the tables in this book offer may find these programs useful; they can either be run on the HP-65 or used as guides to write programs for computers with print-out facilities.

An improved version of *Superellipsoid Radii and Chordal Distances* (1693A) is reproduced below. It handles ellipsoidal and superellipsoidal structures with one circular cross-section. For structures with neither cross-section circular, *GEODESICS III: Superellipsoid Radii* (02693A) solves Equation 13.7 to find vertex radii, and *GEODESICS II: Chord Factors* (02486A) accepts these radii and the vertex coordinates as inputs to give such chord factors as those in Table 15.3.

GEODESICS IV: Face and Dihedral Angles — Spheres (02694A) solves Equations 17.4 through 17.7, using chord factors as inputs. *GEODESICS V: Dihedral and Axial Angles: Non-spheres* combines Equations 17.8 and 17.9 in a single routine. Its inputs are chord factors, radii (from *Geodesics III* or from the superellipsoid program below), and face angles (from *Geodesics IV*). The results are accurate but the multiple inputs make this method of reaching them relatively cumbersome.

GEODESICS X: Spherical Coordinate Rotation (02700A) handles Equations 14.1 and 14.2, either for truncations (Chapter 14) or for rotating an entire system (Chapter 15).

Two programs of general usefulness are given below. One computes chord factors for spherical structures between any two points whose ϕ, θ values are supplied from the tables. It accepts inputs in either decimal degree or arctan notation. The other keeps in storage the expansion and the exponent for an ellipsoidal or superellipsoidal structure with one circular cross-section (see Equation 13.9) and on being given the decimal ϕ, θ values of any two points from a table, computes the two radii and the chord factor of the connecting strut. The three columns in the program listings are step number, keystroke, machine code.

SUPERELLIPSOID RADII AND CHORD FACTORS PROGRAM										
1	RCL 7	3407	26	×	71	51	√	09	76	sin	04
2	RCL 8	3408	27	gx⇄y	3507	52	RCL5	3405	77	×	71
3	g	35	28	RCL8	3408	53	f-1	32	78	+	61
4	y×	05	29	g	35	54	√	09	79	2	02
5	RCL2	3402	30	y×	05	55	+	61	80	×	71
6	f	31	31	+	61	56	f-1	32	81	RCL5	3405
7	TF1	61	32	÷	81	57	SF1	51	82	×	71
8	CLX	44	33	RCL8	3408	58	RCL2	3402	83	RCL6	3406
9	RCL4	3404	34	g	35	59	f	31	84	×	71
10	1	01	35	1/x	04	60	cos	05	85	–	51
11	f-1	32	36	g	35	61	RCL4	3404	86	f	31
12	R→P	01	37	y×	05	62	f	31	87	√	09
13	g	35	38	f	31	63	cos	05	88	RTN	24
14	ABS	06	39	TF1	61	64	×	71	89	LBL	23
15	EEX	43	40	GTO	22	65	RCL1	3401	90	D	14
16	CHS	42	41	B	12	66	RCL3	3403	91	STO2	3302
17	9	09	42	STO5	3305	67	–	51	92	g↓	3508
18	+	61	43	f	31	68	f	31	93	STO1	3301
19	gx⇄y	3507	44	SF1	51	69	cos	05	94	RTN	24
20	gLSTx	3500	45	GTO	22	70	RCL2	3402	95	LBL	23
21	+	61	46	A	11	71	f	31	96	E	15
22	RCL8	3408	47	LBL	23	72	sin	04	97	STO4	3304
23	g	35	48	B	12	73	×	71	98	g↓	3508
24	y×	05	49	STO6	3306	74	RCL4	3404	99	STO3	3303
25	g↑	3509	50	f-1	32	75	f	31	100	RTN	24

Instructions

Store Expansion, *E*, in R7 (key in value and STO 7)
Store Exponent, *n*, in R8 (key in value and STO 8)

These values need not be stored again for subsequent cases.

Data entry: Key ϕ, enter, θ, D. (Stores ϕ and θ of first point in R1
and R2.)

Key ϕ, enter, θ, E. (Stores ϕ and θ of second point in
R3 and R4.)

If a point is reused in a subsequent case its coordinates need not
be stored again.

Run: Press A. After 16 seconds running time, program will halt
and display chord factor. To see radius of first point, press RCL 5.
To see radius of second point press RCL 6.

Comments

Steps 1 to 46 comprise the radius routine, traversed twice. This
solves Equation 13.3, using R←P to obtain $\sin\theta$ and $\cos\theta$ in one
operation. Steps 15-17 generate a particle of "grit"
(0.000000001) which is routinely added to the sine and cosine to
ensure against either value ever being zero: this prevents trouble at
steps 24 and 30, since zero cannot be raised to a power. For the
same reason negative cosines are converted to positive at step 14.
The condition of the flag at step 7 determines whether the radius
routine operates on point 1 or point 2, and its condition at step 39
determines whether the program will reenter the radius routine or
jump to the chord factor routine. Flag is initially unset, set at
step 44 prior to second traverse, again unset at step 57 to prepare
for new case. Chord factor routine (Equation 9.1) runs from step
47 to step 88.

			CHORD FACTORS OF SPHERES								
1	LBL	23	17	sin	04	33	B	12	49	tan	06
2	A	11	18	×	71	34	f⁻¹	32	50	STO3	3303
3	RCL2	3402	19	RCL4	3404	35	tan	06	51	RTN	24
4	f	31	20	f	31	36	STO2	3302	52	LBL	23
5	cos	05	21	sin	04	37	g↓	3508	53	D	14
6	RCL4	3404	22	×	71	38	f⁻¹	32	54	STO2	3302
7	f	31	23	+	61	39	tan	06	55	g↓	3508
8	cos	05	24	2	02	40	STO1	3301	56	STO1	3301
9	×	71	25	×	71	41	RTN	24	57	RTN	24
10	RCL1	3401	26	CHS	42	42	LBL	23	58	LBL	23
11	RCL3	3403	27	2	02	43	C	13	59	E	15
12	−	51	28	+	61	44	f⁻¹	32	60	STO4	3304
13	f	31	29	f	31	45	tan	06	61	g↓	3508
14	cos	05	30	√	09	46	STO4	3304	62	STO3	3303
15	RCL2	3402	31	RTN	24	47	g↓	3508	63	RTN	24
16	f	31	32	LBL	23	48	f⁻¹	32			

Instructions

If data are in "arctan" notation:

First point: arctan ϕ, enter, arctan θ, B (Stored in R1, R2)
Second point: arctan ϕ, enter, arctan θ, C (Stored in R3, R4)

If data are in decimal degree notation:

First point: ϕ, enter, θ, D (Stored in R1, R2)
Second point: ϕ, enter, θ, E (Stored in R3, R4)

B and C keys convert arctan to degrees before storage. Program does not tamper with data, and a point that will be reused in a subsequent case need not be reentered.

To obtain chord factor press A. Running time about 4 1/2 seconds.